Mathematical Modelling of Waves in Multi-Scale Structured Media

MONOGRAPHS AND RESEARCH NOTES IN MATHEMATICS

Series Editors

John A. Burns
Thomas J. Tucker
Miklos Bona
Michael Ruzhansky

Published Titles

Actions and Invariants of Algebraic Groups, Second Edition, Walter Ferrer Santos and Alvaro Rittatore

Analytical Methods for Kolmogorov Equations, Second Edition, Luca Lorenzi

Application of Fuzzy Logic to Social Choice Theory, John N. Mordeson, Davender S. Malik and Terry D. Clark

Blow-up Patterns for Higher-Order: Nonlinear Parabolic, Hyperbolic Dispersion and Schrödinger Equations, Victor A. Galaktionov, Enzo L. Mitidieri, and Stanislav Pohozaev

Bounds for Determinants of Linear Operators and Their Applications, Michael Gil'

Complex Analysis: Conformal Inequalities and the Bieberbach Conjecture, Prem K. Kythe

Computation with Linear Algebraic Groups, Willem Adriaan de Graaf

Computational Aspects of Polynomial Identities: Volume I, Kemer's Theorems, 2nd Edition Alexei Kanel-Belov, Yakov Karasik, and Louis Halle Rowen

A Concise Introduction to Geometric Numerical Integration, Fernando Casas and Sergio Blanes

Cremona Groups and Icosahedron, Ivan Cheltsov and Constantin Shramov

Delay Differential Evolutions Subjected to Nonlocal Initial Conditions Monica-Dana Burlică, Mihai Necula, Daniela Roşu, and Ioan I. Vrabie

Diagram Genus, Generators, and Applications, Alexander Stoimenow

Difference Equations: Theory, Applications and Advanced Topics, Third Edition Ronald E. Mickens

Dictionary of Inequalities, Second Edition, Peter Bullen

Elements of Quasigroup Theory and Applications, Victor Shcherbacov

Finite Element Methods for Eigenvalue Problems, Jiguang Sun and Aihui Zhou

Integration and Cubature Methods: A Geomathematically Oriented Course, Willi Freeden and Martin Gutting

Introduction to Abelian Model Structures and Gorenstein Homological Dimensions Marco A. Pérez

Iterative Methods without Inversion, Anatoly Galperin

Iterative Optimization in Inverse Problems, Charles L. Byrne

Line Integral Methods for Conservative Problems, Luigi Brugnano and Felice Iavernaro

Lineability: The Search for Linearity in Mathematics, Richard M. Aron, Luis Bernal González, Daniel M. Pellegrino, and Juan B. Seoane Sepúlveda

Published Titles Continued

Mathematical Modelling of Waves in Multi-Scale Structured Media, A. B. Movchan, N. V. Movchan, I. S. Jones, and D. J. Colquitt

Modeling and Inverse Problems in the Presence of Uncertainty, H. T. Banks, Shuhua Hu, and W. Clayton Thompson

Monomial Algebras, Second Edition, Rafael H. Villarreal

Noncommutative Deformation Theory, Eivind Eriksen, Olav Arnfinn Laudal, and Arvid Siqveland

Nonlinear Functional Analysis in Banach Spaces and Banach Algebras: Fixed Point Theory Under Weak Topology for Nonlinear Operators and Block Operator Matrices with Applications, Aref Jeribi and Bilel Krichen

Nonlinear Reaction-Diffusion-Convection Equations: Lie and Conditional Symmetry, Exact Solutions and Their Applications, Roman Cherniha, Mykola Serov, and Oleksii Pliukhin

Optimization and Differentiation, Simon Serovajsky

Partial Differential Equations with Variable Exponents: Variational Methods and Qualitative Analysis, Vicenţiu D. Rădulescu and Dušan D. Repovš

A Practical Guide to Geometric Regulation for Distributed Parameter Systems Eugenio Aulisa and David Gilliam

Reconstruction from Integral Data, Victor Palamodov

Signal Processing: A Mathematical Approach, Second Edition, Charles L. Byrne

Sinusoids: Theory and Technological Applications, Prem K. Kythe

Special Integrals of Gradshteyn and Ryzhik: the Proofs – Volume I, Victor H. Moll

Special Integrals of Gradshteyn and Ryzhik: the Proofs – Volume II, Victor H. Moll

Spectral and Scattering Theory for Second-Order Partial Differential Operators, Kiyoshi Mochizuki

Stochastic Cauchy Problems in Infinite Dimensions: Generalized and Regularized Solutions, Irina V. Melnikova

Submanifolds and Holonomy, Second Edition, Jürgen Berndt, Sergio Console, and Carlos Enrique Olmos

Symmetry and Quantum Mechanics, Scott Corry

The Truth Value Algebra of Type-2 Fuzzy Sets: Order Convolutions of Functions on the Unit Interval, John Harding, Carol Walker, and Elbert Walker

Variational-Hemivariational Inequalities with Applications, Mircea Sofonea and Stanislaw Migórski

Willmore Energy and Willmore Conjecture, Magdalena D. Toda

Forthcoming Titles

Groups, Designs, and Linear Algebra, Donald L. Kreher

Handbook of the Tutte Polynomial, Joanna Anthony Ellis-Monaghan and Iain Moffat

Microlocal Analysis on R^n and on NonCompact Manifolds, Sandro Coriasco

Practical Guide to Geometric Regulation for Distributed Parameter Systems, Eugenio Aulisa and David S. Gilliam

MONOGRAPHS AND RESEARCH NOTES IN MATHEMATICS

Mathematical Modelling of Waves in Multi-Scale Structured Media

A. B. Movchan
N. V. Movchan
I. S. Jones
D. J. Colquitt

CRC Press
Taylor & Francis Group
Boca Raton London New York

CRC Press is an imprint of the
Taylor & Francis Group, an **informa** business

A CHAPMAN & HALL BOOK

MATLAB® is a trademark of The MathWorks, Inc. and is used with permission. The MathWorks does not warrant the accuracy of the text or exercises in this book. This book's use or discussion of MATLAB® software or related products does not constitute endorsement or sponsorship by The MathWorks of a particular pedagogical approach or particular use of the MATLAB® software.

CRC Press
Taylor & Francis Group
6000 Broken Sound Parkway NW, Suite 300
Boca Raton, FL 33487-2742

© 2018 by Taylor & Francis Group, LLC
CRC Press is an imprint of Taylor & Francis Group, an Informa business

No claim to original U.S. Government works

International Standard Book Number-13: 978-1-498-78209-8 (Hardback)

This book contains information obtained from authentic and highly regarded sources. Reasonable efforts have been made to publish reliable data and information, but the author and publisher cannot assume responsibility for the validity of all materials or the consequences of their use. The authors and publishers have attempted to trace the copyright holders of all material reproduced in this publication and apologize to copyright holders if permission to publish in this form has not been obtained. If any copyright material has not been acknowledged please write and let us know so we may rectify in any future reprint.

Except as permitted under U.S. Copyright Law, no part of this book may be reprinted, reproduced, transmitted, or utilized in any form by any electronic, mechanical, or other means, now known or hereafter invented, including photocopying, microfilming, and recording, or in any information storage or retrieval system, without written permission from the publishers.

For permission to photocopy or use material electronically from this work, please access www.copyright.com (http://www.copyright.com/) or contact the Copyright Clearance Center, Inc. (CCC), 222 Rosewood Drive, Danvers, MA 01923, 978-750-8400. CCC is a not-for-profit organization that provides licenses and registration for a variety of users. For organizations that have been granted a photocopy license by the CCC, a separate system of payment has been arranged.

Trademark Notice: Product or corporate names may be trademarks or registered trademarks, and are used only for identification and explanation without intent to infringe.

Visit the Taylor & Francis Web site at
http://www.taylorandfrancis.com

and the CRC Press Web site at
http://www.crcpress.com

Printed and bound in Great Britain by
TJ International Ltd, Padstow, Cornwall

Contents

Preface		ix
1	**Introduction**	**1**
	1.1 Bloch–Floquet waves	2
	1.2 Structured interfaces and localisation	3
	1.3 Multi-physics problems and phononic crystal structures	5
	1.4 Designer multi-scale materials	8
	1.5 Dynamic anisotropy and defects in lattice systems	8
	1.6 Models and physical applications in materials science	10
	1.7 Structure of the book	11
2	**Foundations, methods of analysis of waves and analytical approaches to modelling of multi-scale solids**	**13**
	2.1 Wave dispersion	13
	2.2 Bloch–Floquet waves	18
	2.3 Asymptotic lattice approximations	22
	2.4 Transmission and reflection	25
	2.5 Wave localisation and dynamic defect modes	29
	2.6 Dynamic localisation in a biatomic discrete chain	34
	2.7 Asymptotic homogenisation	37
3	**Waves in structured media with thin ligaments and disintegrating junctions**	**49**
	3.1 Structures with undamaged multi-scale resonators	50
	3.2 Singular perturbation analysis of fields in solids with disintegrating junctions	60
	3.3 Structures with damaged multi-scale resonators	72
4	**Dynamic response of elastic lattices and discretised elastic membranes**	**95**
	4.1 Stop-band dynamic Green's functions and exponential localisation	95

4.2	Dynamic anisotropy and localisation near defects	101
4.3	Localisation near cracks/inclusions in a lattice	115

5 Cloaking and channelling of elastic waves in structured solids   141

5.1	A cloak is not a shield	141
5.2	Cloaking as a channelling method for incident waves	142
5.3	Boundary conditions on the interior contour of a cloak	164
5.4	Cloaking in elastic plates	166
5.5	Singular perturbation analysis of an approximate cloak	182

6 Structured interfaces and chiral systems in dynamics of elastic solids   199

6.1	Structured interface as a polarising filter	199
6.2	Vortex-type resonators and chiral polarisers of elastic waves	208
6.3	Discrete structured interface: shielding, negative refraction, and focusing	220

References   229

Index   245

Preface

The study of wave propagation in structured media can be traced as far back as the seventeenth century with the publication of Newton's *Philosophiæ Naturalis Principia Mathematica*. For instance in *Principia*, Newton examined the, now classical, one-dimensional mass-spring lattice system and derived an expression for the speed of propagation of sound waves. Despite being studied for several centuries, wave propagation in multi-scale structured media remains an active, exciting, and challenging area of research.

Indeed, recent developments toward the control of wave propagation in multi-scale solids have led to considerable progress in the development and application of analytical techniques for the modelling of wave propagation in multi-scale structured media. The principal aim of the present text is to provide a unified account of several of these analytical techniques as applied to a collection of fascinating physical problems.

Following the introduction, Chapter 2 summarises several fundamental methods, concepts, and approaches required for the analysis of wave propagation in multi-scale solids. This chapter is designed to furnish the uninitiated reader with a succinct introduction to the primary methods necessary for the material presented later in the text. It is hoped that graduate students and those researchers unfamiliar with the topics covered in the present monograph will find Chapter 2 particularly useful.

The main body of the present text represents a coherent account of a selection of interesting problems tackled over the last 15 years. We were privileged to have worked with wonderful colleagues with whom we have collaborated over the past years, and we are very grateful to them. The book is based on a series of research papers cited throughout the text.

Liverpool, A. B. Movchan
United Kingdom N. V. Movchan
  I. S. Jones
  D. J. Colquitt

COMSOL Multiphysics® is a registered trademark of COMSOL AB.

MATLAB® is a registered trademark of The MathWorks, Inc.
For product information please contact:
The MatWorks, Inc.,
3 Apple Hill Drive,
Natick, MA, 01760-2098 USA,
Tel: 508-647-7000,
Fax: 508-647-7001,
Web: www.mathworks.com

Chapter 1

Introduction

The study of wave propagation in multi-scale structured media has proven a rich area of research, attracting the attention of researchers from a broad range of disciplines including mathematics, physics, and engineering. From a mathematical standpoint, the modelling of dynamic phenomena in structured media is both interesting and challenging and has captivated the applied mathematics community for many decades. In the last century, the field of waves in structured media has undergone a rapid series of developments with both theoretical and experimental advances actively driving the field forward; see, for example, [12, 43, 58, 98, 106, 119–121, 137, 143, 158, 174, 175]. More recently, advances in fabrication methods and materials technology, coupled with the desire and ability to create "designer materials" have led to a significant expansion in experimental research for waves in multi-scale structured media.

Elastic structured media presents a unique challenge in the analysis of wave propagation in multi-scale solids. In the case of electromagnetism, all waves travel at the same speed (the speed of light) in a given medium; the same is true of acoustic waves and water waves. In contrast, elastic bodies generally support the propagation of two different types of wave: pressure waves and shear waves. These two waves travel at different speeds, even in this same medium, and are coupled at boundaries and interfaces; this makes the analysis of wave propagation in structured media, where there are many interfaces and boundaries, singularly difficult. In turn, this unique coupling leads to particularly fascinating dispersion properties and, not only creates several new directions in the mathematical modelling of the dynamic response of multi-scale structured solids, but also allows for the development of many interesting applications in mechanics and engineering.

One particularly interesting area of research relates to *"high-contrast"* structures. In the case of mechanical systems, *"high-contrast"* refers to materials of structures with large spatial variations in density and/or stiffness. Such structures are of particular interest because they can support very low frequency standing waves and localised waveforms. Among the plethora of engineering applications, earthquake protection systems are, understandably, of particular interest and multi-scale systems of resonators can be used to efficiently filter, reflect, polarise, mode convert, and otherwise control elastic waves over a desired frequency range (see, for example, [44, 46]).

In the present book, we consider a broad range of theoretical approaches to the challenging problems of modelling wave propagation and localisation in structured elastic media. In particular, we will focus on multi-scale materials, with a dynamic response that may appear unusual and often counterintuitive; using the rigorous methods developed here, we will show that these, apparently strange, phenomena can be easily understood and are, in fact, entirely natural. Examples of such phenomena include dynamic anisotropy, focussing by flat interfaces and cloaking.

1.1 Bloch–Floquet waves

There are many classical monographs, such as [26], [85], [91], and [187], which address wave propagation in structured media; a topic that has proven to be an exceptionally rich and attractive area of research. Motivated, not only by contemporary theoretical and technological advances, but also by work going back to the seventeenth century [135], there has been a particularly high level of scientific interest in the analysis of systems exhibiting photonic and phononic band gaps for Bloch–Floquet waves. The scholarly interest in the analysis of such phenomena is underlined by the extensive bibliography summarised in [60]. In particular focussing and diffraction, which are well known for optical systems (see, for example, [25] and [145]), can have new and interesting aspects in problems of vector elasticity. A systematic analysis of the scattering of waves in solids with periodic arrays of defects was presented in [35]. For problems of the mathematical theory of elasticity, the paper [166] extended the Rayleigh method [169] to examine the scattering of elastic waves in doubly periodic structures.

For the case of lattice dynamics, *localised primitive waveforms* have been identified [8, 151] and studied in detail. In particular, it was demonstrated that these localised vibrations are associated with stationary points on the dispersion surfaces. For both, time-harmonic and transient excitations, the papers [92] and [93] have analysed the dynamic response of a square lattice, with a particular emphasis on the nature of the caustics, which require careful consideration when applying the method of steepest descent.

For systems of elastic rods, which include periodic arrays of trusses and frames, a generic algorithm for treating the dispersive properties of vector lattices and the design of structures possessing band gaps in desired ranges of frequencies was developed in [108].

1.2 Structured interfaces and localisation

The concept of structured interfaces of finite thickness, which join two continuous regions of an elastic medium, was introduced in [21]; wherein the authors note that the inertial properties of the interface significantly affect the dynamic response of the structure and lead to unusual filtering properties for elastic waves. Extending the work of [21], a structured interface between two continuous domains was employed in [29] to demonstrate the focussing of elastic waves via negative refraction. Such structured interfaces in solid media are sometimes referred to as "flat lenses" for elastic waves. Similar effects have also been demonstrated in acoustics, as shown, for example, in [76]. The effects of focussing and filtering of elastic waves have been extended to entirely discrete structures: a diatomic interface lattice embedded in a monatomic ambient lattice of the same geometry was considered in [49]; and it was shown that, for certain frequencies, the interface lattice acts as a flat elastic lens. Dynamic homogenisation, which also incorporates directional preferences and effects of negative refraction, was addressed in the recent papers [54–56, 139].

Anisotropic inhomogeneous interfaces are useful, for example, for the reduction of stress concentration in solids. Figure 1.1 gives a simple example for a uni-axially loaded static solid containing a set of coated inclusions. The material of the ring-coating is orthotropic and the Young's modulus in the tangential direction within the coating increases toward the centre, similar to [58, 98]. As shown in Figures 1.1c and 1.1d, the stress concentration has been reduced dramatically compared to the cases in Figures 1.1a and 1.1b.

The presence of localised waveforms was previously illustrated for scalar lattices [8, 92, 93, 151]. The resulting anisotropy, diffraction patterns and aberrations are often explained using the dispersion surfaces and slowness contours. Two geometrically identical lattices, with the same elastic stiffness may provide different dynamic responses due to different distributions of inertia and embedded resonators of different kind. In statics such lattices, which we denote as type I and type II, may be geometrically identical. However, embedding an interface composed of a lattice of type I into an ambient lattice of type II will result in highly fascinating dynamic features. Although in statics we would have a uniform lattice which is homogenous, its dynamic response may exhibit the total reflection of waves by the interface, as well as negative refraction and focussing.

Imperfect interfaces may influence the overall properties of the homogenised composites. In particular, there are interesting examples concerning the notion of neutral inclusions and, in a static setting, these were studied in [14, 23, 100, 122].

Dynamic composite structures containing coated inclusions have been analysed in [160, 161], where the notion of neutrality was linked to the effective refractive index in the long-wave limit; the coated inclusions were considered

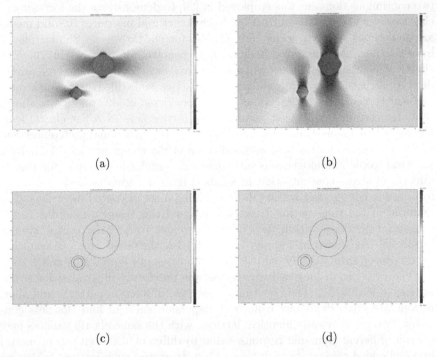

FIGURE 1.1: Panels (a) and (b) show the stress σ_{11} and σ_{22} field plots around an inclusion in a uni-axially loaded solid; (c) and (d) panels show the stress σ_{11} and σ_{22} field plots around an inclusion surrounded by an orthotropic inhomogeneous coating in the same uni-axially loaded solid. The stress concentration has been dramatically reduced.

as neutral if the effective refractive index of the composite medium was the same as that of the homogenous solid without inclusions. In these papers, the elastic coating around an inclusion was considered as anisotropic, and it was demonstrated that the geometrical and physical parameters of the coatings could be chosen to match the effective refractive index of the doubly periodic array of coated inclusions with the refractive index of the unperturbed homogeneous medium. Acoustic band gaps for waves in media with neutral inclusions were considered in [77].

In statics one can introduce so-called *imperfect interfaces* in order to describe high-contrast coatings with elastic properties that are significantly different from those of the ambient elastic medium. In the context of homogenisation and bounds on effective moduli and neutrality, static models of composites containing coated inclusions have been analysed in [13, 15, 78, 79, 101]; including both stiff and soft, as well as highly anisotropic imperfect interfaces in two-dimensional elasticity and problems of torsion.

A model of *non-local structured interfaces* was developed in [21] for static and dynamic problems in elastic composite media. Special features, highlighted in that paper, include strong anisotropy and a finite-width interface. In particular, for dynamic problems the structured interface may possess trapped waveforms, which enhance transmission. The papers [16–18] studied new static models of non-local structured interfaces, and addressed evaluation of their effective mechanical properties and their influence on the elastic stress concentration around defects. Structured interfaces possess inertia, and hence, tractions may become discontinuous across such interfaces in dynamic problems of wave propagation.

The interaction between periodic discrete and continuous systems was studied in [21, 29] wherein the existance of "slow waves" was noted, and the transmission properties of a finite-width dynamic structured interface was analysed leading to unusual dynamic effects such as negative refraction.

1.3 Multi-physics problems and phononic crystal structures

The interaction of electromagnetic waves with photonic crystal structures has been the subject of rapid development in the last decade. In particular, cloaking of electromagnetic waves has been analysed both experimentally and theoretically; some of the earliest investigations in this area of wave studies were reported in [58, 174]. Being natural and well-studied for problems of electromagnetism and acoustics, the concept of wave cloaking was also addressed for other classes of physical problems, which include linear water waves [63] and flexural vibration of elastic plates [64]. The equations of vector elasticity

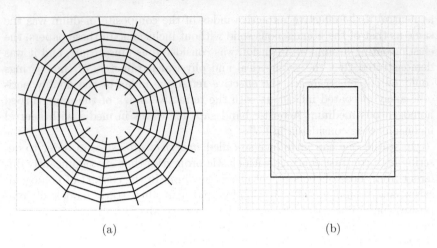

FIGURE 1.2: Structured interface coating: (a) the radial anisotropic structure, (b) the square anisotropic coating.

bring additional challenges, compared to the Maxwell system or the equations describing flexural waves in elastic structures; the tensor structure of the governing equations of mathematical elasticity incorporates the notion of shear. As a result, pressure and shear waves propagate with different speeds and couple through the boundary conditions. This makes the modelling of elastic metamaterial systems, which possess cloaking properties, difficult as outlined, for example, in [119, 143].

The dynamic response of vector elastic lattice systems, interacting with waves, has been the subject of classical investigations, as discussed, for example in [91], as well as [49, 177–179]. We note substantial differences between scalar problems involving the vibration of systems of harmonic springs and vector problems of elasticity, referring to elastic rods and beams, which connect a system of finite solids or point masses. The notion of shear stress and shear strain is absent in scalar lattice systems and there is no such analogue in models of electromagnetism. The misconception of predictability of dynamic properties of vector elastic lattices is sometimes based on an intuitive extrapolation of results available for the scalar systems or for problems of electromagnetism. As illustrated in [49], results observed for micro-polar elastic lattice structures do not always follow from the simpler scalar physical models or intuitive assumptions.

The papers [99, 129, 160–162, 166, 190] have addressed the problems of dispersion of Bloch–Floquet waves and phononic band gaps for elastic waves in doubly periodic solids containing arrays of voids or inclusions. Transmission problems for waves interacting with arrays of elastic gratings of inclusions were studied in [163, 164], including a comparative analysis of the filtering proper-

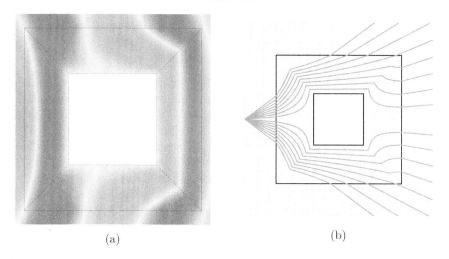

(a) (b)

FIGURE 1.3: Wave-cloaking of a square scatterer: (a) a distorted wave front within the cloaking layer; (b) the ray diagram showing the distorted metrics in the cloaking region.

ties of elastic waves in doubly periodic media and the transmission properties for the corresponding singly periodic gratings consisting of circular inclusions.

An elegantly designed elastic coating, with anisotropic inertial properties, has been proposed in [123] to model an elastic "invisibility cloak", which takes an incident wave around a scatterer. The paper [28] shows an alternative design of an elastic cloaking coating which engages a micro-polar inhomogenous composite structure. Structured interfaces, and their discrete approximations, can be effectively used to construct dynamic invisibility cloaks, with two examples shown in Figure 1.2, which illustrate a radially symmetric structure (part(a)) and a square structured interface (part (b)). The radially symmetric wave cloaks are discussed, for example, in [43, 58, 98], and the square cloak has been modelled in [48].

The papers [30, 39, 40] have addressed propagation of elastic waves, their dispersion properties, and their interaction with defects in a two-dimensional structured medium endowed with micro-rotations associated with spinning gyroscopes embedded into the lattice system. The analysis of gyroscopic motion of an individual spinner was incorporated into the system of conservation of linear and angular momenta within the lattice systems. Consequently, it has led to a special design of chiral media, which are characterised by "handedness" in their microstructures. Such media possess special properties related to shielding, polarisation and filtering of elastic waves.

1.4 Designer multi-scale materials

The term *"metamaterials"* is often used for multi-scale materials, which are engineered to possess specially designed properties that are not found in nature. Examples of interesting designs include structures which exhibit negative refraction [176], filtering, polarisation and focussing of waves [62,74,118]. In particular, metamaterial wave cloaking structures have been designed and implemented in many physical applications; the example in Figure 1.3 shows a distorted wave within a square inhomogeneous and anisotropic cloaking interface and the ray diagram illustrating "diversion" of an incident wave [48].

Another term, "hyperbolic metamaterials", is used for a special class of structured multi-scale media, which exhibit a strong dynamic anisotropy accompanied by star-shaped wave forms. Such materials were designed and studied in problems of optics [165], as well as in acoustics and linear elasticity [8,50,92,94,131,151,170]. In particular, in the framework of time-harmonic fields, the waves in hyperbolic metamaterials are modulated by functions, which satisfy hyperbolic partial differential equations.

The paper [52] introduced the concept of "parabolic metamaterials" as multi-scale structures which support uni-directionally localised waveforms. This work also demonstrated interesting connections with the Dirac points on the dispersion surfaces for Bloch–Floquet waves in a special frequency range. We also note that Dirac cones on the dispersion surfaces occur as the result of degeneracies in the dispersion equations characterising Bloch–Floquet waves in photonic [117], phononic [42], and platonic crystals [112].

The dynamic homogenisation scheme developed and demonstrated in [5,6,47,55,56,103] is also useful in the description of hyperbolic and parabolic metamaterials. Classical two-scale asymptotic homogenisation approaches for microstructured media (see, for example, [12,154]) usually involve the study of a class of model problems on an elementary cell and can be used effectively in the long-wave quasi-static regime. In contrast, the approach of dynamic homogenisation addresses perturbations away from known resonances and may be used to study dynamic anisotropy and localisation of waves in dispersive media.

1.5 Dynamic anisotropy and defects in lattice systems

Classical applications in the theory of defects in crystals and dislocations follow from the fundamental work [105], where explicit closed-form solutions were derived for a heterogeneous lattice system when two distant particles of different masses are interchanged. The envelope function-based perturbation approach was developed in [59, 102] for analysis of waveguides in photonic crystal structures. In the latter case, an array of cylindrical inclusions embedded into an ambient medium represents a waveguide, where the frequencies of the guided modes are close to the band edge of the unperturbed doubly periodic system.

Periodic lattices may exhibit very different behaviour in dynamics compared to their static response. In particular, dynamic anisotropy and different classes of localisation have been identified in periodic lattices, as outlined in [8, 50, 92, 94, 131, 151, 170]. In some frequency regimes, corresponding to neighbourhoods of resonances, hyperbolic metamaterials can be viewed as multi-scale structures, where Bloch–Floquet waves exhibit behaviour associated with locally hyperbolic dispersion surfaces in the neighbourhood of certain resonances. Strong spatial localisation may occur where wave propagation is permitted only along directions associated with the principal curvatures of the hyperbolic dispersion surface. Important connections between the dispersion properties of Bloch–Floquet waves in a periodic lattice and the solutions of forced problems are discussed in [50].

The dynamic lattice Green's function describes the vibration of a lattice with a single-point defect, or applied point force. Green's functions have been studied in [107] for two-dimensional square lattices in pass band regimes. The paper [131] addressed continuous and discrete models for exponentially localised waveforms, with various forcing or defect types. Localised waveforms have been identified for the cases when the forcing frequency and/or the natural frequency of the defect are placed in the band gap. Such defect modes can be linked to the stop-band Green's functions.

Using an asymptotic approach, the papers [20, 68] considered the effect of a pre-stress on the propagation of flexural waves through an elastic beam on a Winkler foundation. Special attention was devoted to dynamically localised waveforms and to control the position of band gaps via pre-stress. It was found that a tensile pre-stress can increase the frequency at which a particular band gap occurs.

The monograph [178] presents a detailed discussion of applications for dynamic lattice problems involving cracks modelled as semi-infinite faults, advancing in two-dimensional periodic elastic lattices. For a structured interface and a crack propagating with an average constant speed through a square lattice, localised modes were analysed in [124]. The article [138] presented the study of a semi-infinite dynamic crack in a non-uniform elastic lattice. The crack stability was analysed and a connection has been established between the unstable crack growth and the energy of waves emanating from the crack-tip in the steady-state regime.

A comprehensive mathematical theory developing an asymptotic analysis of fields in multi-structures is presented in the monograph [90], where boundary layers near junctions are shown to have special importance in both static problems as well as in problems of time-harmonic vibrations.

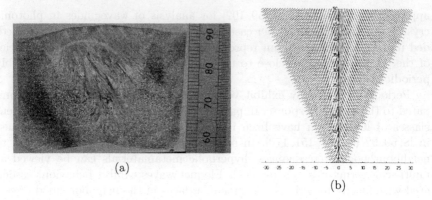

FIGURE 1.4: Panel (a) shows a photograph of a typical weld which exhibits the characteristic columnar grain alignments. Panel (b) is a plot of the grain orientations generated by a scanning electron microscope from an actual weld. *Images courtesy of Amec Foster Wheeler plc.*

1.6 Models and physical applications in materials science

Studies of the propagation of elastic waves in inhomogeneous and anisotropic materials are often motivated by the requirement for effective ultrasonic non-destructive examination of strongly anisotropic steel welds and other components. In particular, an inhomogeneity may arise when the casting or weld is formed and the metallic crystals align such that their symmetry axes are parallel to the maximal thermal gradient. Figure 1.4 shows a typical weld and grain map, illustrating the inhomogeneity.

In some applications, ultrasonic probes and sensors are used to detect defects in welded structures, and this in turn may require measurements of the scattered fields. A numerical ray-tracing model was developed in [146] to account for the inhomogeneous and anisotropic nature of the weld. The paper [140] presented an investigation into the effect of beam distortion in anisotropic, but homogeneous materials. Further studies [2, 186] suggested that the weld may be modelled as a material with constant density and crystalline moduli, but with the orientations of the crystals smoothly varying with position. In addition, the paper [144] developed a ray theory for elastic waves propagating through the inhomogeneous and anisotropic medium, with the emphasis on the special case of SH-waves.

Applications of the Bloch–Floquet wave theory to the study of the dynamic localisation in welds has been illustrated in [51], which has surpassed the traditional ray-tracing approaches and revealed new features linked to dispersion and localisation of elastic waves in a weld with a granular structure. In par-

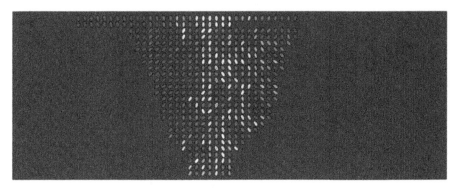

FIGURE 1.5: An illustration of the effect of grain structure on the path of ultrasonic waves through granular welds. Here we plot the magnitude of the dilatation.

ticular, the authors of [51] developed an analytical and computational model capable of using data taken from actual welds (such as the grain structure shown in Figure 1.4) to analyse the propagation of ultrasonic waves through granular welds. Figure 1.5 shows an illustrative computation highlighting the localisation and complex wave behaviour exhibited in such welds.

1.7 Structure of the book

This text is based on recent theoretical advances leading to modelling of dynamic anisotropy, localisation and design of multi-scale heterogeneous materials, which possess unusual dynamic properties such as negative refraction, filtering, and focussing of waves by flat interfaces as well as the creating of invisibility cloaks. Some background material and a theoretical introduction are also included in the book. Some of the key topics presented here include Green's functions and localised waveforms, dynamic response of dynamic elastic structures with thin ligaments, dynamic anisotropy and localisation in lattice systems.

Furthermore, the state-of-the-art is presented for models of hyperbolic and parabolic modes and uni-directional or star-shaped wave forms, disintegrating elastic solids and effective junction conditions for multi-structures in dynamics. Multi-resonator systems, which also include tuneable thermo-elastic dynamic systems, are studied using asymptotic approximations, combined with numerical simulations.

Asymptotic analysis is also used here to study continuous and discrete media, dynamic inclusions and cracks in lattices versus Bloch–Floquet waves, platonic crystals and localised defect modes, chiral systems and models of vibrating gyro-lattices, approximate cloaking in lattice structures, as well as scattering reduction in structured plates.

Modelling foundations and methods of analysis of waves in multi-scale media are discussed in Chapter 2. Chapter 3 includes models of waves in so-called "disintegrating solids", which are multi-scale structures with thin ligaments and disintegrating junctions. Dynamic anisotropy, wave localisation and defect waveforms in lattice structures are presented in Chapter 4. Chapter 5 deals with the topic of cloaking via the introduction of specially designed multi-scale heterogeneous coatings. Finally, in Chapter 6, the reader will find models of structured interfaces and special chiral media developed for multi-scale elastic systems.

The book is aimed at a wide audience, including applied mathematicians, physicists and engineers. It may be especially useful to research students who would like to study dispersion and localisation of waves in structured media.

Enjoy reading!

Chapter 2

Foundations, methods of analysis of waves and analytical approaches to modelling of multi-scale solids

This preliminary chapter introduces some fundamental concepts and methods required for the analysis of wave propagation in multi-scale solids. We begin in Section 2.1 by introducing the concept of dispersion, initially for the classical example of linear water waves, then followed by one-dimensional mass-spring chains. The elements of Bloch–Floquet theory for infinite periodic media are summarised in Section 2.2, followed by a discussion of asymptotic approximations of high-contrast continua by discrete lattices in Section 2.3. We then proceed to study a one-dimensional transmission problem where two dissimilar lattices are joined by a *"structured interface"* in Section 2.4; we also introduce the notion of the transmission matrix. An important connection is established between formally distinct Bloch–Floquet problem and the transmission problem.

In Sections 2.5 and 2.6 we move on to problems for defects in otherwise periodic media and analyse the associated localised waves. Finally, in Section 2.7, we briefly summarise the approach of Craster et al. [47, 55] for the asymptotic homogenisation of periodic media at finite frequencies, which has proven extremely useful in the analysis of dynamic problems in multi-scale media.

The present chapter is intended to provide the uninitiated reader with a concise introduction to the principal methods required for the material presented in later chapters. The interested reader is also referred to the references provided for further information. In contrast, the reader familiar with this area of research may consider proceeding directly to Chapter 3.

2.1 Wave dispersion

This introductory section discusses some classical examples of dispersive waves, i.e. waves whose speeds are different for different frequencies. At the beginning we refer to linear water wave theory, which is well known in models of fluid flow as discussed, for example, in [24]. Another example, included in Section 2.2, is a one-dimensional lattice system, known from classical texts, such as [85] and [26]. The dispersion relations, that is the equations relating the permissible values of frequency and wavenumber, are written in explicit form and formation of a stop band, i.e. ranges of frequency with which no waves can propagate, is discussed for heterogeneous systems. An asymptotic model is given for a high-contrast stratified structure in Section 2.2. Following [128] we discuss the lattice approximation of such a system, which is capable of reproducing the dynamic response of the heterogeneous continuum system in the low frequency range.

2.1.1 Elementary considerations for linear water waves

One of the straightforward examples of dispersion is based on the linear theory of waves propagating along the surface of an incompressible inviscid fluid of uniform density. For convenience, we call such a fluid "water". The continuity equation and the equation of motion for the velocity $\mathbf{u}$ and pressure p have the form [3]

$$\nabla \cdot \mathbf{u} = 0, \tag{2.1}$$

$$\frac{\partial \mathbf{u}}{\partial t} + (\mathbf{u} \cdot \nabla)\mathbf{u} + \frac{1}{\rho}\nabla p = \mathbf{F}, \tag{2.2}$$

where $\mathbf{F}$ is the body force density, ρ is the mass density, and t is time. In particular, if the body force represents gravity we have $\mathbf{F} = -g\mathbf{e}^{(3)}$, and

$$\mathbf{F} = \nabla \Xi \quad \text{with} \quad \Xi = -gx_3, \tag{2.3}$$

where g is a positive constant, normalised gravitational acceleration, the upward vertical x_3-axis is orthogonal to the unperturbed water surface, and $\mathbf{e}^{(3)}$ is the unit vector along this axis.

Assuming that the fluid flow is irrotational and using the notation ϕ for the velocity potential we have $\mathbf{u} = \nabla \phi$. Hence according to (2.1) the function ϕ is harmonic

$$\nabla^2 \phi = 0, \tag{2.4}$$

within the fluid. The non-linear term in (2.2) becomes

$$\{(\mathbf{u} \cdot \nabla)\mathbf{u}\}_i = \sum_j u_j \frac{\partial u_i}{\partial x_j} = \sum_j \frac{\partial \phi}{\partial x_j} \frac{\partial^2 \phi}{\partial x_i \partial x_j} = \frac{1}{2} \frac{\partial}{\partial x_i}(|\nabla \phi|^2). \tag{2.5}$$

Thus, (2.2) may be written as

$$\nabla \left(\frac{\partial \phi}{\partial t} + \frac{1}{2}|\nabla \phi|^2 + \frac{p}{\rho} + gx_3 \right) = 0,$$

which leads to Bernoulli's equation

$$\frac{\partial \phi}{\partial t} + \frac{1}{2}|\nabla \phi|^2 + \frac{p}{\rho} + gx_3 = f(t), \tag{2.6}$$

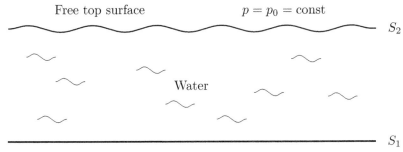

FIGURE 2.1: Surface water waves in a flow of finite depth.

where $f(t)$ is a function of t only.

The boundary conditions are set at the bottom and upper surfaces S_1 and S_2, respectively, as shown in Figure 2.1. It is assumed that the bottom surface S_1 of the channel is fixed and hence the normal component of the velocity is zero, that is,

$$\mathbf{u} \cdot \mathbf{n} = 0 \quad \text{on} \quad S_1. \tag{2.7}$$

It is also assumed that the upper surface S_2 is characterised by the equation

$$x_3 = \zeta(x_1, x_2, t),$$

and there is no variation in pressure on S_2:

$$p = p_0 = \text{const} \quad \text{on} \quad S_2.$$

Hence, according to (2.6) on the free surface we have

$$\frac{\partial \phi}{\partial t} + \frac{1}{2}|\nabla \phi|^2 + \frac{p_0}{\rho} + g\zeta(x_1, x_2, t) = f(t), \quad \text{as } x_3 = \zeta(x_1, x_2, t). \tag{2.8}$$

This is accompanied by the identity

$$u_3 = \frac{dx_3}{dt} = \frac{d\zeta}{dt} = \frac{\partial \zeta}{\partial t} + u_1 \frac{\partial \zeta}{\partial x_1} + u_2 \frac{\partial \zeta}{\partial x_2} \quad \text{on} \quad x_3 = \zeta(x_1, x_2, t). \tag{2.9}$$

In the linear approximation, we simplify the problem by addressing the case when $|\mathbf{u}|$, the surface fluctuations, and the derivatives $\partial \zeta/\partial x_1$ and $\partial \zeta/\partial x_2$ are all small.

In particular, if the unperturbed surface is $x_3 = h = \text{const}$, then for the free surface we can write

$$\phi(\mathbf{x}, t) = \phi(x_1, x_2, \zeta(x_1, x_2, t), t) =$$
$$= \phi(x_1, x_2, h, t) + (\zeta - h)\frac{\partial \phi}{\partial x_3}(x_1, x_2, h, t) + \ldots$$

To the leading order approximation, the relations (2.8) and (2.9) are set on the unperturbed surface $x_3 = h$. The second-order term $\frac{1}{2}|\nabla\phi|^2$ is neglected, and Equation (2.8) becomes

$$\frac{\partial \phi}{\partial t} + \left(\frac{p_0}{\rho} - f(t) + gh\right) + g(\zeta - h) = 0.$$

Since the term $\left(\frac{p_0}{\rho} - f(t) + gh\right)$ is a function of t only, it does not give any contribution to $\mathbf{u}$, and may be "absorbed" into $\frac{\partial \phi}{\partial t}$. Thus, the velocity potential ϕ can be chosen in such a way that

$$\frac{\partial \phi}{\partial t} + g(\zeta - h) = 0 \quad \text{on} \quad x_3 = h. \tag{2.10}$$

By neglecting the second-order terms $(\partial \phi/\partial x_1)(\partial \zeta/\partial x_1)$, $(\partial \phi/\partial x_2)(\partial \zeta/\partial x_2)$ in (2.9), involving the partial derivatives of ζ and ϕ, we deduce that to the leading order

$$\frac{\partial \phi}{\partial x_3} = \frac{\partial \zeta}{\partial t} \quad \text{on} \quad x_3 = h, \tag{2.11}$$

and therefore Equations (2.10), (2.11) yield

$$\frac{\partial^2 \phi}{\partial t^2} + g\frac{\partial \phi}{\partial x_3} = 0 \quad \text{on} \quad x_3 = h. \tag{2.12}$$

Thus, in the "ocean" with the flat bottom surface $x_3 = 0$ and the unperturbed upper surface $x_3 = h$, the linearised formulation for the velocity potential ϕ is comprised of Laplace's equation (2.4) within the fluid layer $0 < x_3 < h$, the boundary condition $\mathbf{n} \cdot \nabla \phi = 0$ at the fixed bottom surface $x_3 = 0$, and the "free surface" boundary condition (2.12). Note that this formulation involves the second-order time derivative of the velocity potential on the upper free surface. In turn, after evaluation of ϕ, the "wave profile" on the free upper surface is defined by (2.10).

2.1.2 Dispersion equation

Consider a time-harmonic water wave of radian frequency ω, with the straight front perpendicular to the x_1–axis. We seek solutions independent of x_2 with the velocity potential

$$\phi = \phi(x_1, x_3, t). \tag{2.13}$$

It is convenient to work with the complex potential Ψ of the form

$$\Psi = W(x_3)e^{-i(\omega t - kx_1)},$$

and assume that $\phi = \text{Re}\Psi$. Considering a harmonic solution (i.e. $\nabla^2 \Psi = 0$) we obtain

$$W'' - k^2 W = 0,$$

which yields
$$W = C_1 \cosh(kx_3) + C_2 \sinh(kx_3).$$

The boundary condition at the bottom surface ($x_3 = 0$) leads to $W'(0) = 0$, and hence $C_2 = 0$ and

$$\Psi = C_1 \cosh(kx_3) e^{-i(\omega t - kx_1)}. \tag{2.14}$$

In turn, the free surface condition

$$\left. \left(\frac{\partial^2 \Psi}{\partial t^2} + g \frac{\partial \Psi}{\partial x_3} \right) \right|_{x_3 = h} = 0 \tag{2.15}$$

becomes

$$\{-\omega^2 C_1 \cosh(kh) + gkC_1 \sinh(kh)\} e^{-i(\omega t - kx_1)} = 0,$$

which simplifies to the form

$$gk \tanh(kh) = \omega^2. \tag{2.16}$$

Equation (2.16) relates the wave number k and the radian frequency ω, and thus the [phase] wave speed $c = \frac{\omega}{k}$ is in general frequency dependent. This shows that the water surface waves are *dispersive*. The relation (2.16) is called the *dispersion equation*.

2.1.2.1 Asymptotics: deep and shallow water waves

We would like to note two important particular cases, which are

(a) Deep water, when the non-dimensional quantity kh is large ($kh \gg 1$), and hence $\tanh(kh) \simeq 1$. Thus to leading order the dispersion equation (2.16) yields

$$\omega^2 = gk.$$

In this case, the wave speed increases when the frequency decreases, i.e. $c = g\omega^{-1}$, which is fully consistent with the physical observation that low frequency waves in the ocean propagate faster than the high frequency ripples.

(b) Shallow water, when $kh \ll 1$, and hence $\tanh(kh) \sim kh$. In this case, to leading order the dispersion equation (2.16) takes the form

$$\omega^2 = gk^2 h,$$

which implies $c = \sqrt{gh}$. In the framework of this approximation, the wave speed in shallow water is frequency-independent and hence the shallow water waves can be treated as non-dispersive.

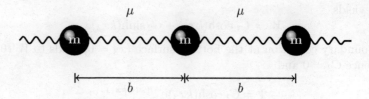

FIGURE 2.2: Periodic chain of springs and masses.

It is noted that the dispersion of surface water waves in the above illustrative example is linked to the boundary condition (2.12), which involves the second-order time derivative of the velocity potential.

The phenomenon of wave dispersion can also be observed in simple periodic structures. Conventionally, dispersion diagrams are used to describe the change in frequency with the wave number, which will be illustrated in the next section.

2.2 Bloch–Floquet waves

A Bloch–Floquet wave (sometimes referred to as a Bloch wave or a Floquet wave) in a periodic system is an object discussed in classical textbooks, such as [85] and [26]. Here we show elementary examples referring to the time-harmonic motion of a one-dimensional periodic lattice, consisting of point masses (rigid particles) connected by massless springs. The sketch of such a system is shown in Figure 2.2.

We start by considering a mass-spring chain where all particles have the same mass, m, and are connected to their nearest neighbours by springs of stiffness μ and length b.

Let u_n be the longitudinal displacement from equilibrium of the nth node within the chain, with the interatomic spacing b. Then the equations of motion take the form

$$m\ddot{u}_n = \mu(u_{n+1} + u_{n-1} - 2u_n) \quad \text{with } n \text{ being integer.} \qquad (2.17)$$

If the motion is time harmonic then $u_n = U_n \exp(-i\omega t)$, and therefore

$$-m\omega^2 U_n = \mu(U_{n+1} + U_{n-1} - 2U_n). \qquad (2.18)$$

In the case of travelling waves, we have

$$U_n = U e^{inkb}, \quad \text{with } U = \text{const}, \qquad (2.19)$$

and the equation of motion becomes

$$-\omega^2 m U = 2\mu U(\cos(kb) - 1).$$

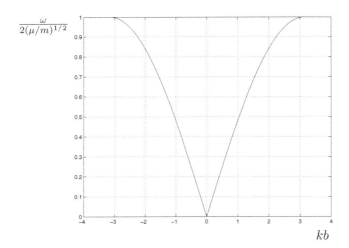

FIGURE 2.3: The dispersion curves $\omega = \omega(k)$ which characterises the dispersion properties of Bloch–Floquet waves propagating in the periodic spring-mass system shown in Figure 2.2.

The quantity k is said to be the Bloch parameter, and a solution u_n of (2.17), which satisfies the condition

$$u_{s+n} = u_s e^{inkb},$$

is referred to as the *Bloch–Floquet wave* [26, 85].

A non-trivial solution U_n of (2.18) and (2.19) exists, provided ω and k satisfy the following dispersion equation:

$$\omega^2 - \frac{2\mu}{m}(1 - \cos(kb)) = 0,$$

whose non-negative solution is

$$\omega = 2\sqrt{\frac{\mu}{m}} \left| \sin\left(\frac{kb}{2}\right) \right|. \tag{2.20}$$

The corresponding dispersion curve, representing a periodic function of k, with the period $2\pi/b$, is shown in Figure 2.3. The interval $(-\pi/b, \pi/b)$ is known as the *Brillouin zone*. Conventionally the dispersion diagrams are displayed for the values of k from the Brillouin zone.

2.2.1 Standing waves

It is now convenient to refer to the general case of an arbitrary interatomic separation b within the lattice. In particular, taking $b \to 0$ one obtains the continuum limit. We note that

$$U_{n+1}/U_n = e^{ikb},$$

and within the Brillouin zone we have $k_{max} = \pi/b$. In the continuum limit, as $b \to 0$, we deduce $k_{max} \to \infty$.

The transmission velocity of a wave packet is called the *group velocity*, and it is given as

$$v_g = \frac{d\omega}{dk} = \sqrt{\frac{\mu}{m}} b \cos\left(\frac{kb}{2}\right), 0 < k < \pi/b.$$

On the boundaries $k = \pm\pi/b$ of the Brillouin zone, we have $v_g = 0$; the solution U_n represents a *standing wave* with zero net transmission velocity, and $U_n = (-1)^n U$.

In many physical applications, it is useful to have asymptotic approximations corresponding to the *long-wave limit*. In particular, this will give the slope of the curve $\omega = \omega(k)$ as $k \to 0+$, which is also referred to as the effective group velocity v_{eff}. This is often used in *homogenisation* approximations of wave phenomena in structured media. When $kb \ll 1$, we have

$$\omega \simeq \sqrt{\frac{\mu}{m}} |kb|.$$

Therefore, we deduce that $v_{\text{eff}} = (\mu/m)^{1/2} b$, and, of course, v_{eff} is frequency-independent in this limit.

2.2.2 Stop bands

For the next example, also outlined in [26] and [85], we consider the one-dimensional periodic lattice consisting of two types of particles, of different masses. This gives an excellent illustration of filtering properties of structured media.

Compared to the case of a uniform spring-mass chain, the new periodic system has two types of masses, m_1 and m_2, and springs of normalised stiffness μ (see Figure 2.4). In this case the elementary cell of the periodic lattice includes two different particles, with displacements u_n and v_n, and one needs two equations of motion as follows

$$m_1 \frac{d^2 u_n}{dt^2} = \mu(v_n + v_{n-1} - 2u_n), \quad m_2 \frac{d^2 v_n}{dt^2} = \mu(u_{n+1} + u_n - 2v_n). \quad (2.21)$$

Let D be the size of the elementary cell, which is equal to the distance between the nearest particles of the same mass. The displacements at the nodes of the nth cell in a travelling wave are given by

$$u_n = U e^{i(nkD - \omega t)}, \quad v_n = V e^{i(nkD - \omega t)},$$

where k is the wave number.

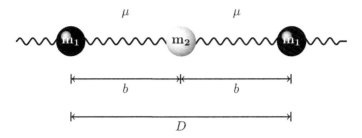

FIGURE 2.4: Periodic spring-mass chain containing two types of masses.

Using the equations of motion, we deduce

$$\omega^2 \mu^{-1} m_1 U + V(1 + e^{-ikD}) - 2U = 0,$$

$$\omega^2 \mu^{-1} m_2 V + U(1 + e^{ikD}) - 2V = 0.$$

This is a system of homogenous linear algebraic equations with respect to U and V. It has a non-trivial solution if and only if the determinant of the matrix of this system vanishes, i.e.

$$\frac{m_1 m_2}{\mu^2} \omega^4 - \frac{2(m_1 + m_2)}{\mu} \omega^2 + 2(1 - \cos(kD)) = 0. \qquad (2.22)$$

This is the *dispersion equation* providing the relationship between ω and k. For the present system, this equation has explicit solutions of the form

$$\omega^2 = \mu \frac{m_1 + m_2 \pm \sqrt{(m_1 + m_2)^2 - 2m_1 m_2(1 - \cos(kD))}}{m_1 m_2}. \qquad (2.23)$$

Equation (2.23) implies that the dispersion diagram, corresponding to Equation (2.22) has two branches. They are conventionally referred to as the "acoustic branch" (corresponding to the sign "−" in (2.23)) and the "optical branch" (corresponding to the sign "+" in (2.23)), as shown in Figure 2.5. It is noted that a similar technique is used to obtain dispersion equations for wave propagation in discrete lattices in more general periodic systems.

We note that for the non-homogeneous system, when $m_1 \neq m_2$, there is a non-zero separation between the dispersion curves, a *stop band*. The width of the stop band can be computed by evaluating the frequencies at the end points of the Brillouin zone. At the boundary of the Brillouin zone, when $k = \pm \pi/D$, the roots of the dispersion equation are defined by

$$\omega^2 = \frac{\mu}{m_1 m_2} \{m_1 + m_2 \pm |m_1 - m_2|\}.$$

Assuming, without loss of generality that $m_1 < m_2$, one can deduce that the width of the band gap is equal to $\sqrt{2\mu/m_1} - \sqrt{2\mu/m_2}$. If we fix m_2 and let

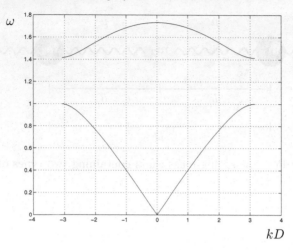

FIGURE 2.5: Dispersion curves for the Bloch–Floquet waves in the spring-mass structure, which includes two types of masses; here we use $m_1 = 1, m_2 = 2, \mu = 1$.

the contrast parameter $r = m_2/m_1$ increase, then the width of the stop band will increase.

No propagating wave exists within the interval of frequencies $(\sqrt{2\mu/m_2}, \sqrt{2\mu/m_1})$. If a forced motion is initiated at a frequency within that interval, we get an evanescent wave.

From the dispersion diagram in Figure 2.5, we can identify the *high frequency stop band* defined as the interval $(\omega_*, +\infty)$, where $\omega_*^2 = 2\mu(m_1^{-1} + m_2^{-1})$ is the upper limit for frequencies corresponding to the optical branch of the dispersion diagram, obtained as $k \to 0$.

For the biatomic lattice system, sinusoidal Bloch–Floquet waves exist only within the frequency intervals $[0, \sqrt{2\mu/m_2}]$ and $\left[\sqrt{2\mu/m_1}, \sqrt{2\mu(m_1^{-1} + m_2^{-1})}\right]$, which are referred to as the *pass bands*. The position and the width of the *pass bands* can be controlled by changing the stiffness of springs and the masses of particles.

2.3 Asymptotic lattice approximations

We have demonstrated in the previous section that the entire pass-band set, incorporating all pass-band intervals for a discrete periodic lattice, is bounded. In contrast, a continuum system would support propagation of high-frequency waves. For high-contrast periodic structures, it is sometimes possible

Foundations and methods

to construct a "lattice approximation", which describes the dispersion properties of Bloch–Floquet waves in the low frequency range. An example of such a structure is discussed in this section, which is based on the paper [132].

Instead of a discrete system of point masses connected by massless springs, we consider here a one-dimensional periodic array of elastic rods of alternating stiffnesses μ_j, $j = 1, 2$, and non-zero mass density. For convenience, it is assumed that the linear mass density ρ is the same for all elements of this structure. The elementary cell contains two types of elastic rods. We use the notations $S_1^{(n)} = (-b + nD, nD)$, $S_2^{(n)} = (nD, a + nD)$, where n is integer, and $D = a + b$ is the total size of the elementary cell.

Assuming that the waves are time-harmonic, of the radian frequency ω, we deduce that the amplitudes of the longitudinal displacements $U_j(x)$, $j = 1, 2$, satisfy the equations

$$\mu_j U_j'' + \omega^2 \rho U_j = 0, \quad x \in S_j^{(n)}, \ j = 1, 2. \tag{2.24}$$

The ideal contact conditions at the interface between two neighbouring rods imply continuity of both displacements and tractions, such that

$$U_1 = U_2, \quad \mu_1 U_1' = \mu_2 U_2'. \tag{2.25}$$

The solution is sought in the class of Bloch–Floquet waves, for which we can write

$$U_j(x + D) = e^{iKD} U_j(x), \ j = 1, 2, \ |K| < \pi/D. \tag{2.26}$$

The general solution of (2.24) is

$$U_j = A_j e^{ik_j x} + B_j e^{-ik_j x}, \quad x \in S_j^{(n)}, \ j = 1, 2,$$

where A_j and B_j are constant coefficients, and $k_j(\omega) = \omega\sqrt{\rho/\mu_j}$, are linear functions of ω. The interface conditions (2.25) and the Bloch–Floquet condition (2.26), written for displacements within the elementary cell, lead to a homogeneous system of linear algebraic equations with respect to A_j and B_j:

$$\mathbf{Q}(K, \omega) \begin{pmatrix} A_1 \\ B_1 \\ A_2 \\ B_2 \end{pmatrix} = 0, \tag{2.27}$$

where

$$\mathbf{Q}(K, \omega) = \begin{pmatrix} 1 & 1 & -1 & -1 \\ \mu_1 k_1 & -\mu_1 k_1 & -\mu_2 k_2 & \mu_2 k_2 \\ -e^{i(KD-k_1 b)} & -e^{i(KD+k_1 b)} & e^{ik_2 a} & e^{-ik_2 a} \\ k_1 e^{i(KD-k_1 b)} & -k_1 e^{i(KD+k_1 b)} & -\frac{\mu_2}{\mu_1} k_2 e^{ik_2 a} & \frac{\mu_2}{\mu_1} k_2 e^{-ik_2 a} \end{pmatrix}.$$

$$\tag{2.28}$$

This system has a non-trivial solution, provided

$$\det \mathbf{Q}(K,\omega) = 0, \qquad (2.29)$$

which relates the radian frequency ω to the Bloch parameter K and hence gives the dispersion equation for Bloch–Floquet waves propagating along the periodic structure. It is convenient to introduce the notation $\varepsilon = \mu_1/\mu_2$, and in this case $k_2 = k_1\sqrt{\varepsilon}$. Then (2.29) leads to

$$(\varepsilon+1)\sin(k_1 b)\sin(k_1 a\sqrt{\varepsilon}) - 2\sqrt{\varepsilon}(\cos(k_1 b)\cos(k_1 a\sqrt{\varepsilon}) - \cos(KD)) = 0. \quad (2.30)$$

Assume that $\varepsilon \ll 1$ and $k_1 b \ll 1$. This implies that the structure has a high-contrast in the stiffness of its phases; also the length b of one of the elastic sections in the elementary cell is assumed to be relatively small. Then the corresponding trigonometric parts in the left-hand side of (2.30) can be expanded into power series, and after truncation we obtain a polynomial expression in powers of ω.

Retaining the terms up to ω^4, we obtain the *approximate dispersion equation*

$$\omega^4 \mathcal{P} - \omega^2 \mathcal{Q} + 2(1 - \cos(KD)) = 0, \qquad (2.31)$$

where the quantities $\mathcal{P}$ and $\mathcal{Q}$ are positive and have the form

$$\mathcal{P} = \frac{\rho^2}{12/\mu_1^2}(\varepsilon a^2 + 2\varepsilon ba + b^2)(\varepsilon a^2 + b^2 + 2ba), \quad \mathcal{Q} = \frac{\rho}{\mu_1}(a+b)(\varepsilon a + b).$$

Here, Equation (2.31) has the same structure as (2.22) derived for a discrete system, which includes a one-dimensional array of masses m_1, m_2 connected by elastic springs of stiffness μ. Hence a discrete lattice, consisting of springs and masses, can be used to approximate a high-contrast periodic continuous system to obtain its dynamic response within the low frequency range.

This discrete model is characterised by the two parameters m_1/μ and m_2/μ, and Equations (2.31) and (2.22) become the same when

$$\mathcal{P} = m_1 m_2/\mu^2, \quad \mathcal{Q} = 2(m_1 + m_2)/\mu.$$

We note that the transcendental dispersion equation characterising waves in the bi-material continuum system has an infinite number of solutions, whereas the lattice approximation covers only the first two dispersion curves adjacent to the origin. An illustrative diagram is included in Figure 2.6, which shows the solutions of the dispersion equation (2.30) for the waves in the high-contrast two-phase system; this figure also shows the region where the lattice approximation applies.

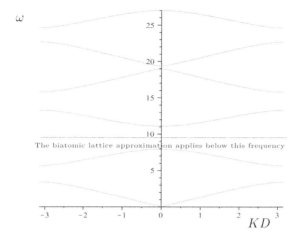

FIGURE 2.6: Dispersion diagram for Bloch–Floquet waves in the high-contrast bi-material periodic system. The parameter values, used in the computations, are $D = 1, b = 0.5, \varepsilon = 0.1$.

2.4 Transmission and reflection

Here, we describe a one-dimensional "structured interface", which incorporates a finite number of elastic rods aligned with the x-axis and joined together sequentially. The rods themselves may have different lengths, densities and elastic stiffnesses. A wave propagating in the positive direction of the x-axis interacts with the interface, and some of the energy is reflected whereas the remaining energy is transmitted through the interface. The percentage of the reflected energy is frequency dependent. For the case when there is a repeating pattern, and the size of such an interface increases to infinity to form a periodic structure, it is appropriate to make a connection with the analysis of Bloch–Floquet waves and their dispersion properties in the periodic system, as explained in [31, 97]. However, when the thickness of the interface structure is not large, the analysis follows a different pattern. The emphasis is on trapped modes, which may exist within the structured interface and which may modify and enhance its transmission properties. The transmission matrix technique, as in [97], is an efficient tool which we outline next.

2.4.1 Transmission matrix

The displacement amplitude u for the time-harmonic motion of radian frequency ω satisfies the equation

$$u''(x) + \frac{\rho}{\mu}\omega^2 u(x) = 0, \tag{2.32}$$

where ρ and μ are the mass density and the stiffness constant of a single rod, respectively. The waveform can be expressed in terms of complex amplitudes A and B

$$u(x) = A\exp(ikx) + B\exp(-ikx), \tag{2.33}$$

with

$$k = \omega c, \quad c = \sqrt{\frac{\rho}{\mu}}.$$

Consider a one-phase interface located between $x = x_0$ and $x = x_1 = x_0 + d$. Then displacements and tractions are related by

$$\begin{pmatrix} u(x_1) \\ \mu\dfrac{\partial u}{\partial x}(x_1) \end{pmatrix} = \mathbf{M} \begin{pmatrix} u(x_0) \\ \mu\dfrac{\partial u}{\partial x}(x_0) \end{pmatrix},$$

$$= \begin{pmatrix} m_{11} & m_{12} \\ m_{21} & m_{22} \end{pmatrix} \begin{pmatrix} u(x_0) \\ \mu\dfrac{\partial u}{\partial x}(x_0) \end{pmatrix}, \tag{2.34}$$

where, according to (2.33), we have

$$\mathbf{M} = \begin{pmatrix} \cos\delta & \dfrac{\sin\delta}{Q} \\ -Q\sin\delta & \cos\delta \end{pmatrix}, \tag{2.35}$$

with $Q = \mu k$, and $\delta = kd$ being the *phase increment*. The matrix $\mathbf{M}$ in (2.35) is said to be the *transmission matrix* and its eigenvalues are

$$\cos\delta \pm \sqrt{\cos^2\delta - 1} = \exp(\pm i\delta). \tag{2.36}$$

Along the same lines, we now analyse a *discrete interface* consisting of two springs of stiffness γ and a mass m. The transmission matrix $\mathbf{M}^{(d)}$ of the discrete interface is written in the form

$$\mathbf{M}^{(d)} = \begin{pmatrix} \dfrac{\gamma - m\omega^2}{\gamma} & \dfrac{2\gamma - m\omega^2}{\gamma^2} \\ -m\omega^2 & \dfrac{\gamma - m\omega^2}{\gamma} \end{pmatrix}. \tag{2.37}$$

Foundations and methods

The eigenvalues of the transmission matrix (2.37) are

$$1 - \frac{m}{\gamma}\omega^2 \pm \sqrt{\frac{m}{\gamma}\omega^2 \left(\frac{m}{\gamma}\omega^2 - 2\right)}, \qquad (2.38)$$

so that the dependence on ω is algebraic for the structured discrete interface, in contrast with the continuous one-phase interface (see (2.36)). It is also noted that the eigenvalues (2.38) are complex for sufficiently small ω and real when ω is sufficiently large, in particular, when $\omega^2 > 2\gamma/m$.

If several interface regions, continuous or discrete, are placed together to form an interface stack, then the transmission matrix is obtained by the appropriate multiplication of the transmission matrices of individual layers within the stack, as discussed in [31].

2.4.2 Reflected and transmitted energy

We consider an interface stack with the overall 2×2 transmission matrix $\mathbf{M} = (m_{ij})$ separating two elastic media (for the one-dimensional case, we have two semi-infinite elastic rods located along the intervals $x < x_0$ and $x > x_1$), with the density ρ_- and stiffness μ_- on the left from the interface and the density ρ_+ and stiffness μ_+ on the right from the interface. The *incident* and *reflected* waves on the left of the interface admit the representation

$$u(x) = A_I \exp(ik_- x) + A_R \exp(-ik_- x), \qquad (2.39)$$

where A_I and A_R are the *incident* and *reflection amplitudes*, while the transmitted wave on the right of the interface is given by

$$u(x) = \sqrt{\frac{\mu_-}{\mu_+}} A_T \exp(ik_+ x), \qquad (2.40)$$

with A_T denoting the *transmission amplitude*, and $k_\pm$ indicating the corresponding wave numbers in the regions outside the interface.

According to the definition of the transmission matrix outlined in the previous section (c.f. Equation (2.34)), the coefficients A_R, A_T and A_I satisfy the equation

$$\sqrt{\frac{\mu_-}{\mu_+}} \begin{pmatrix} A_T e^{ik_+ x_1} \\ iQ_+ A_T e^{ik_+ x_1} \end{pmatrix} = \mathbf{M} \begin{pmatrix} A_I e^{ik_- x_0} + A_R e^{-ik_- x_0} \\ iQ_-(A_I e^{ik_- x_0} - A_R e^{-ik_- x_0}) \end{pmatrix}, \qquad (2.41)$$

where $Q_\pm = \mu_\pm k_\pm$. In turn, the amplitudes A_R and A_T of the reflected and transmitted waves can be written (see [31]) in the form

$$\bar{A}_R = \frac{Q_- Q_+ m_{12} + m_{21} + i(Q_- m_{22} - Q_+ m_{11})}{Q_- Q_+ m_{12} - m_{21} + i(Q_- m_{22} + Q_+ m_{11})} e^{2ik_- x_0},$$

$$\bar{A}_T = \sqrt{\frac{\mu_+}{\mu_-}} \frac{2iQ_-}{Q_- Q_+ m_{12} - m_{21} + i(Q_- m_{22} + Q_+ m_{11})} e^{i(k_- x_0 - k_+ x_1)}, \qquad (2.42)$$

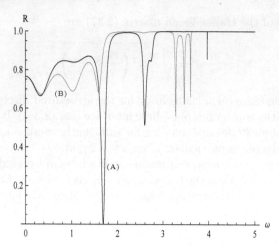

FIGURE 2.7: Normalised reflected energy for the interface consisting of 4 cells, mM-Mm-mM-mM (curve A), versus the normalised reflected energy for the interface mM-mM-mM-mM (curve B).

where we have used the normalised quantities $\bar{A}_R = A_R/A_I$, $\bar{A}_T = A_T/A_I$.

According to the standard procedure (see, for example, [31, 97]), the normalised *reflected* R and *transmitted* T *energies* are given by

$$R = |\bar{A}_R|^2 = \frac{(Q_-Q_+m_{12} + m_{21})^2 + (Q_-m_{22} - Q_+m_{11})^2}{(Q_-Q_+m_{12} - m_{21})^2 + (Q_-m_{22} + Q_+m_{11})^2},$$

$$T = |\bar{A}_T|^2 \frac{Q_+}{Q_-} = \frac{4Q_-Q_+}{(Q_-Q_+m_{12} - m_{21})^2 + (Q_-m_{22} + Q_+m_{11})^2}.$$

(2.43)

Here, $T + R = 1$, so that the total transmission corresponds to the case of $R = 0$, $T = 1$, and total reflection corresponds to $R = 1$, $T = 0$.

2.4.3 Defect modes and enhanced transmission

There is a connection between the transmission properties of a finite width stack and dispersion properties of Bloch–Floquet waves in an infinite doubly periodic planar medium.

In particular, if a full stop band is identified for Bloch–Floquet waves, then the finite-width stack built of the same material is expected to reflect the bulk of the energy within this stop band. However, the presence of defects in the stack may alter its transmission properties. This is illustrated here for a simple one-dimensional example.

We consider the one-dimensional problem involving the spring-mass interface, described above. In the computational example below, the discrete interface is composed of masses $m_1 = 387.3\,\text{kg}$ and $m_2 = 3m_1$ connected by identical springs of stiffness $\gamma = 3873\,\text{N} \cdot \text{m}^{-1}$. The dispersion of waves, for the two-mass lattice system, is explained in Section 2.2.2. In particular, the dispersion diagram includes the frequency intervals $(\sqrt{2\gamma/m_2}, \sqrt{2\gamma/m_1})$ and $(\sqrt{2\gamma(m_1^{-1} + m_2^{-1})}, +\infty)$, where no sinusoidal waves can propagate (stop bands). It is anticipated that within the stop-band frequency interval a finite sub-structure $m_1m_2 - m_1m_2 - m_1m_2 - m_1m_2$ of this periodic system can be used as a structured interface, so that a substantial proportion of the energy of the incident wave would be reflected by the interface.

The plots of the normalised reflected energy as a function of the frequency are shown in Figure 2.7 where a biatomic lattice interface is inserted between two different infinite continuous media. The continuous media on the left and on the right from the interface have the elastic impedances $\sqrt{\rho_-\mu_-} = 100\,\text{Pa} \cdot \text{m}^{-1}$ and $\sqrt{\rho_+\mu_+} = 1500\,\text{Pa} \cdot \text{m}^{-1}$, respectively.

In particular, Figure 2.7 includes two curves representing the normalised reflected energy for the interfaces $m_1m_2 - m_2m_1 - m_1m_2 - m_1m_2$ and $m_1m_2 - m_1m_2 - m_1m_2 - m_1m_2$. The interchange of the order of two neighbouring masses creates a "defect mode", which leads to an enhanced transmission. This is shown as a sharp reduction in the reflected energy, corresponding to the curve (A) in Figure 2.7.

2.5 Wave localisation and dynamic defect modes

Stop bands and localisation for waves in structured media, as well as related applications in electromagnetism and telecommunications, have been of interest to physicists for nearly 100 years. In the beginning of the twentieth century, Campbell developed a theory of electric wave filters and used the idea of stop-band localisation [37]. Several important applications of Green's functions in models of lattice dynamics are also outlined in [106].

In particular, lattices with defects have interesting applications in solid state physics, due to the appearance of so-called "localised defect waveforms", which are studied in a series of related papers [113–116] as well as the paper [104]. Defect vibration modes in two-component discrete systems which have special filtering properties, were described in [9]. Dynamic lattice Green's functions for periodic structures in regimes, which correspond to propagating waves, were studied in [107]. One of the important features of dynamic periodic lattices is a strong anisotropy, influenced by both the frequency and Bloch wavevector. In this context, the so-called "star-shaped" waveforms in lattices have been studied in [8, 50, 92, 93, 170].

The present section is based on the paper [131], which developed the concept of the "band-gap Green's function". It is mainly used to describe exponentially localised vibrations, that may occur in both discrete and continuous systems. This theory has also described localised waveforms associated with a perturbation of mass within the elastic body, which is outlined in the text below.

2.5.1 Localisation of waves in a flexural beam on an elastic foundation

Here, we show a simple example of a localised waveform in a homogeneous flexural elastic beam, which is placed on an elastic foundation. There is no discrete structure involved in this problem; nevertheless, a low frequency stop band is identified.

In the time-harmonic regime, the equation of motion for the displacement amplitude u in a flexural beam on an elastic foundation is given by (see, for example, [182])

$$Du^{(IV)}(x) + (\varkappa - \varrho\omega^2)u(x) = 0 , \qquad (2.44)$$

where ω is the radian frequency, and x is the coordinate along the beam; the flexural rigidity D and the mass density ρ are positive constants. We use the notation $\varkappa$ for the stiffness of the elastic foundation, and $\Omega = \omega^2$. Assuming that

$$u(x) = e^{ikx},$$

we obtain the dispersion equation

$$\omega^2 = \frac{Dk^4 + \varkappa}{\varrho}. \qquad (2.45)$$

Propagating waves correspond to the real k solutions of (2.45); complex solutions describe localised vibration modes. It follows that there is the lower bound ω_- for the sinusoidal wave frequency, below which only evanescent waveforms exist. Namely, when $k = 0$,

$$\omega_-^2 = \Omega_- = \varkappa/\varrho. \qquad (2.46)$$

When $0 < \Omega < \Omega_-$, and when a unit force is applied at the origin, the flexural displacement function takes the form

$$u(x) = \frac{1}{8D\lambda^3}e^{-\lambda|x|}(\cos\lambda x + \sin\lambda|x|), \quad \lambda = \left(\frac{\varkappa - \varrho\Omega}{4D}\right)^{1/4}. \qquad (2.47)$$

Consequently, the localised mode exists for $\Omega < \varkappa/\varrho$ if a concentrated mass M is attached to the elastic beam. The mass-frequency relation is

$$M = \frac{1}{\varrho\omega^2}\sqrt{8}(\varkappa - \varrho\omega^2)^{3/4}D^{1/4}. \qquad (2.48)$$

2.5.2 Flexural plate on an elastic foundation: localisation

In a similar way, a localised waveform can be created in an elastic flexural plate. In the time-harmonic regime, for a plate on an elastic foundation, the equation of motion implies

$$D\Delta^2 u(x,y) + (\varkappa - \varrho\Omega)u(x,y) = 0. \qquad (2.49)$$

Repeating the procedure used in Section 2.5.1 above, but this time searching for solutions of the form $u(x,y) = e^{i(k_x x + k_y y)}$, we find that the same lower bound (2.46) for the sinusoidal wave frequency applies here. For evanescent waves, the displacement produced by the unit source placed at the origin, and evaluated at $x = y = 0$, is given by

$$u(0,0) = \frac{1}{8\sqrt{D(\varkappa - \varrho\Omega)}}. \qquad (2.50)$$

A localised waveform, corresponding to a radian frequency $\omega = \sqrt{\Omega}$, $0 < \Omega < \Omega_-$ can be created by adding a point mass. For a given Ω within the low-frequency stop band, the required value of an additional concentrated mass is

$$M = \frac{8\sqrt{D(\varkappa - \varrho\Omega)}}{\varrho\Omega}. \qquad (2.51)$$

Using this formula, one can create exponentially localised waveforms associated with time-harmonic flexural waves in a continuous elastic plate supported by an elastic foundation.

2.5.3 Wave localisation in a non-local material

In a classical linear elastic homogeneous material, the waves are not dispersive, and a sinusoidal waveform exists for any positive frequency, i.e. there are no stop bands. On the other hand, a change in the constitutive law may induce wave localisation.

An example concerning waves in a one-dimensional elastic rod provides an illustration. The simplest constitutive law for a homogeneous linear elastic material is

$$\sigma = Eu'(x) = E\delta'(x) * u(x) = E\delta(x) * u'(x), \qquad (2.52)$$

where $\delta(x)$ represents the Dirac delta function, and $*$ denotes the operation of convolution.

One can also apply an exponential cut-off procedure, replacing the delta-function by the regularised function δ_α:

$$\delta_\alpha(x) = \frac{\alpha}{2} e^{-\alpha|x|}, \qquad (2.53)$$

$$\delta'_\alpha(x) = -\frac{\alpha^2}{2} e^{-\alpha|x|} \operatorname{sign} x. \qquad (2.54)$$

It is noted that $\delta_\alpha(x) \to \delta(x), \delta'_\alpha(x) \to \delta'(x)$, as $\alpha \to \infty$, in the sense of distributions. Using the Fourier transform

$$[U]^F(k) = \int_{-\infty}^{\infty} U(x)e^{ixk}dx,$$

we can write

$$[\delta_\alpha(x)]^F = \frac{\alpha^2}{\alpha^2 + k^2}, \quad [\delta'_\alpha(x)]^F = -\frac{\alpha^2 ik}{\alpha^2 + k^2}, \qquad (2.55)$$

and instead of the Fourier transform of the classical equation for waves in a rod

$$[Eu'(x)]' + \varrho\Omega u(x) = 0 \qquad (2.56)$$

with constant modulus of elasticity E and density ϱ, we obtain a new equation of the form

$$\left(\frac{E\alpha^2 k^2}{\alpha^2 + k^2} - \varrho\Omega\right) u^F(k) = 0. \qquad (2.57)$$

The solutions of the dispersion equation for the string with non-local interaction have the upper bound $\Omega = \Omega_+ = c^2\alpha^2$, where $c = \sqrt{E/\varrho}$. It is the regularisation of the delta function (2.53), which is responsible for the non-local nature of the interaction.

For $\Omega > \Omega_+$, we consider a generalised solution of the problem, corresponding to a time-harmonic point force applied to the elastic rod at the origin. The Fourier transform of the displacement and the displacement itself are given by

$$u^F(k) = -\frac{k^2 + \alpha^2}{(\varrho\Omega - E\alpha^2)k^2 + \varrho\alpha^2\Omega}, \qquad (2.58)$$

$$u(x) = -\frac{1}{\varrho\Omega(1 - c^2\alpha^2/\Omega)}\left(\delta(x) - \frac{a^2 - \alpha^2}{a}e^{-a|x|}\right), \qquad (2.59)$$

where

$$a = \frac{\alpha}{\sqrt{1 - c^2\alpha^2/\Omega}}, \quad \Omega > \alpha^2. \qquad (2.60)$$

This generalised solution implies that the additional point mass, required for the localised oscillations of the radian frequency $\omega = \sqrt{\Omega}$, is

$$M = -\varrho(1 - c^2\alpha^2/\Omega) > -\varrho \quad (x=0), \quad M + \rho > 0. \qquad (2.61)$$

A regularisation, preserving the condition $\varrho + M > 0$, can be achieved by replacing the forcing term $\delta(x)$ and the corresponding displacement (2.59) as follows:

$$\delta(x) \to \delta(x) * H(b - |x|) = H(b - |x|),$$

$$u(x) \to u(x) * H(b - |x|), \quad b \leq \frac{1}{2a}. \qquad (2.62)$$

Subject to the mass density change within the region $|x| < b$, the displacement is also regularised.

2.5.4 Waves in a chain of particles on an elastic foundation

Here we modify the problem, already discussed in Section 2.2 for a chain of point masses connected by springs. Now, the particles are placed on an elastic foundation, so that in the time-harmonic regime of radian frequency ω, following the application of the discrete Fourier transform, the equation of motion leads to

$$\left(\varkappa + 2(1 - \cos k) - \omega^2\right) u^F(k) = 1. \qquad (2.63)$$

As in the previous sections, the notation $\varkappa$ is used for the stiffness of the elastic foundation, and $\Omega = \omega^2$. The mass of particles, the elastic rigidity of springs and the separation between neighbouring particles have been normalised.

The sinusoidal waveforms occur at the values of ω such that $\omega_- < \omega < \omega_+$, where

$$\omega_+^2 = \Omega_+ = 4 + \varkappa \quad \text{and} \quad \omega_-^2 = \Omega_- = \varkappa. \qquad (2.64)$$

Two band gaps are identified, $\omega > \omega_+$ and $0 < \omega < \omega_-$, where no sinusoidal wave can propagate.

2.5.4.1 Higher frequency band gap

When $\omega > \omega_+$, the displacements produced by the time-harmonic force, of unit amplitude, applied at the origin, become exponentially localised

$$u(m) = -\frac{2^{-|m|} \left(\sqrt{(\Omega - \varkappa)^2 - 4(\Omega - \varkappa)} - (\Omega - \varkappa - 2)\right)^{|m|}}{\sqrt{(\Omega - \varkappa)^2 - 4(\Omega - \varkappa)}}, \qquad (2.65)$$

and such a waveform also can be obtained by adding a mass M (if M is negative, then the mass is subtracted) to the central node at the origin. The required mass appears to be negative

$$M = \frac{1}{\Omega u(0)} = -\frac{\sqrt{(\Omega - \varkappa)^2 - 4(\Omega - \varkappa)}}{\Omega} \quad \text{where } \omega^2 = \Omega > 4 + \varkappa, \qquad (2.66)$$

which implies that a reduction of one of nodal masses, in such a way that $0 < 1 + M < 1$, leads to exponentially localised oscillations when $\omega > \sqrt{4 + \varkappa}$.

2.5.4.2 Lower frequency band gap

When the time-harmonic point force is applied at the origin at a radian frequency ω, such that, $0 < \omega < \omega_-$, we also observe an exponential localisation

$$u(m) = \frac{2^{-|m|} \left(\varkappa - \Omega + 2 - \sqrt{(\varkappa - \Omega)^2 + 4(\varkappa - \Omega)}\right)^{|m|}}{\sqrt{(\varkappa - \Omega)^2 + 4(\varkappa - \Omega)}}. \qquad (2.67)$$

Such a waveform can be obtained by adding the mass M to the central node at the origin, where

$$M = \frac{1}{\Omega u(0)} = \frac{\sqrt{(\varkappa - \Omega)^2 + 4(\varkappa - \Omega)}}{\Omega} > 0 \quad \text{with } 0 < \Omega < \varkappa. \quad (2.68)$$

The localised oscillations within the lower frequency band gap occur only if $\varkappa > 0$, and the lower frequency band gap disappears when $\varkappa = 0$.

2.6 Dynamic localisation in a biatomic discrete chain

We continue the theme of wave localisation in a mass-spring system by allowing for a non-uniform mass distribution along the chain. Namely, we consider a perturbation problem for a biatomic chain. For the periodic case, which was already discussed in Section 2.2.2, the band gaps (or stop bands) on the dispersion diagram (Figure 2.5) have been identified. Here, we construct the localised waveforms at frequencies corresponding to these band gaps.

For a biatomic chain, where point masses m_1 and m_2 are connected by massless springs, we assume that $m_1 > m_2$. The equations of motion, written for two neighbouring nodal masses, are

$$\begin{aligned} m_1 \ddot{u}_{1,n} &= u_{2,n} + u_{2,n-1} - 2u_{1,n}, \\ m_2 \ddot{u}_{2,n} &= u_{1,n} + u_{1,n+1} - 2u_{2,n}. \end{aligned} \quad (2.69)$$

For a wave, characterised by the following nodal displacements

$$u_{1,n} = u_{1,0} e^{ikn}, \quad u_{2,n} = u_{2,0} e^{ikn}, \quad (2.70)$$

the two periodic dispersion curves are defined by the equations

$$\Omega = \frac{m_1 + m_2}{m_1 m_2} \pm \sqrt{\left(\frac{m_1 + m_2}{m_1 m_2}\right)^2 - \frac{4 \sin^2(k/2)}{m_1 m_2}}, \quad (2.71)$$

where $\Omega = \omega^2$, with ω being the radian frequency.

As shown in Figure 2.5 there are two band gaps on the dispersion diagram. The first, finite width, band gap is defined by

$$\Omega_- = \frac{2}{m_1} < \Omega < \frac{2}{m_2} = \Omega_+, \quad (2.72)$$

and the second band gap is semi-infinite with lower bound

$$\Omega > \frac{2}{m_1} + \frac{2}{m_2}. \quad (2.73)$$

Within the band gaps, the waveforms are exponentially localised.

2.6.1 Point forces applied to the central cell

Assume that time-harmonic point forces P_1 and P_2 are applied to the masses m_1 and m_2, respectively, in the central cell ($n = 0$). Using the notations $u_{1,n}$ and $u_{2,n}$ for the amplitudes of vibrations, and $\delta_{l,n}$ for the Kronecker delta, the equations of motion yield

$$-m_1 \Omega u_{1,n} = u_{2,n} + u_{2,n-1} - 2u_{1,n} + P_1 \delta_{0,n},$$
$$-m_2 \Omega u_{2,n} = u_{1,n} + u_{1,n+1} - 2u_{2,n} + P_2 \delta_{0,n}. \qquad (2.74)$$

Furthermore, the spatial discrete Fourier transform leads to

$$u_1^F = \frac{1}{Q}\left(-P_1(m_2 \Omega - 2) + P_2(1 + e^{ik})\right), \qquad (2.75)$$

$$u_2^F = \frac{1}{Q}\left(-P_2(m_1 \Omega - 2) + P_1(1 + e^{-ik})\right), \qquad (2.76)$$

where

$$Q = (m_1 \Omega - 2)(m_2 \Omega - 2) - 2(1 + \cos k). \qquad (2.77)$$

We also note that $Q < 0$ for the first, finite-width, band gap, and $Q > 0$ for the second band, which has an infinite extent.

2.6.2 Localised vibration modes within the finite band gap

Assuming that

$$\frac{2}{m_1} < \Omega < \frac{2}{m_2}, \qquad (2.78)$$

we obtain the displacements at nodal points by applying the inverse Fourier transform, which gives

$$u_{1,m} = -\frac{1}{\sqrt{Q_0^2 - 4Q_0}}\left\{\left(-P_1(m_2\Omega - 2) + P_2\right)\lambda^{|m|} + P_2 \lambda^{|m-1|}\right\},$$

$$u_{2,m} = -\frac{1}{\sqrt{Q_0^2 - 4Q_0}}\left\{\left(-P_2(m_1\Omega - 2) + P_1\right)\lambda^{|m|} + P_1 \lambda^{|m+1|}\right\}, \qquad (2.79)$$

where

$$\lambda = \frac{1}{2}\left(\sqrt{Q_0^2 - 4Q_0} + Q_0 - 2\right), \qquad Q_0 = (m_1\Omega - 2)(m_2\Omega - 2) < 0.$$

We also note that the quantity λ is such that $-1 < \lambda < 0$, and hence the above solution is exponentially localised. In particular, the displacements of nodal points in the central cell, associated with the origin, are

$$u_{1,0} = \frac{P_1(m_2\Omega - 2) - (1/2)P_2\left(Q_0 + \sqrt{Q_0^2 - 4Q_0}\right)}{\sqrt{Q_0^2 - 4Q_0}},$$

$$u_{2,0} = \frac{P_2(m_1\Omega - 2) - (1/2)P_1\left(Q_0 + \sqrt{Q_0^2 - 4Q_0}\right)}{\sqrt{Q_0^2 - 4Q_0}}. \quad (2.80)$$

2.6.3 Perturbation of mass

Now we consider the case of free waves in the absence of external forcing, and introduce masses $m_1 + M_1$ and $m_2 + M_2$ for nodal points in the central cell ($n = 0$). The perturbation terms M_1, M_2 can be positive or negative; if M_j is negative, it implies that the mass of the corresponding particle has been reduced.

Formally, the representation (2.79) can be used with $P_1 = M_1\Omega u_1$, $P_2 = M_2\Omega u_2$, and the condition of the existence of a non-trivial solution has the form

$$0 = \left(\sqrt{Q_0^2 - 4Q_0} - M_1\Omega(m_2\Omega - 2)\right)$$

$$\times \left(\sqrt{Q_0^2 - 4Q_0} - M_2\Omega(m_1\Omega - 2)\right)$$

$$- \frac{M_1 M_2 \omega^4}{4}\left(\sqrt{Q_0^2 - 4Q_0} + Q_0\right)^2, \quad (2.81)$$

where $2/m_1 < \Omega < 2/m_2$. This equation provides the choice of the perturbations M_1, M_2 for the masses in the central cell in order to obtain an exponentially localised waveform of the radian frequency $\omega = \sqrt{\Omega}$ from the first band gap.

In particular, the previous equation (2.81) implies that when $M_2 = 0$, the other perturbation of mass is given by

$$M_1 = -\frac{1}{\Omega}\sqrt{\frac{m_1\Omega - 2}{2 - m_2\Omega}}\sqrt{4 - Q_0},$$

$$= -\frac{2}{\Omega(1 + r)}\sqrt{\frac{r\Omega - 1 - r}{1 + r - \Omega}}\sqrt{(1 + r)^2 + (r\Omega - 1 - r)(1 + r - \Omega)}. \quad (2.82)$$

On the other hand, when $M_1 = 0$, the required value for M_2 is

$$M_2 = \frac{1}{\Omega(1 + r)}\sqrt{\frac{2 - m_2\Omega}{m_1\Omega - 2}}\sqrt{4 - Q_0},$$

$$= \frac{2}{\Omega(1 + r)}\sqrt{\frac{1 + r - \Omega}{r\Omega - 1 - r}}\sqrt{(1 + r)^2 + (r\Omega - 1 - r)(1 + r - \Omega)}. \quad (2.83)$$

Here, $r = m_1/m_2 > 1$ is the mass ratio and the normalisation is introduced in such a way that $m_1 + m_2 = 2$, so that the corresponding homogeneous lattice is the same as that in Section 2.5.4. In this case

$$Q_0 = 4\left(\frac{\Omega}{1+r} - 1\right)\left(\frac{r\Omega}{1+r} - 1\right). \tag{2.84}$$

The analytical formulae (2.82) and (2.83) suggest that a localised defect mode can be initiated by a small variation of one of the masses in the central cell. Namely, a localised mode will appear near the lower edge of the band gap, i.e. $\Omega \to 2m_1^{-1} + 0$, when $M_2 = 0$ and $M_1 \sim -C\sqrt{m_1\Omega - 2}$, with C being a positive constant. We note that the localised mode near the lower edge of the band gap cannot be created by a small variation of m_2 with $M_1 = 0$.

On the contrary, to obtain a localised vibration at the frequency close to the upper edge of the band gap, i.e. $\Omega \to 2m_2^{-1} - 0$, it is sufficient to increase the smaller mass m_2, so that $M_2 \sim C\sqrt{2 - m_2\Omega}$, $C > 0$, while $M_1 = 0$. The localised mode near the upper edge of the band gap cannot be created by a small variation of m_1 while $M_2 = 0$. It is also noted that the band gap region cannot be covered by creating waveforms corresponding to a finite increase of the mass m_2 while $M_1 = 0$. However, it has been demonstrated in [131] that by a simultaneous finite variation of both masses, m_1 and m_2, localised waveforms can be obtained over an entire frequency range within the finite band gap.

2.7 Asymptotic homogenisation

Essentially, asymptotic homogenisation involves the scaling of either spatial or temporal variables by a small parameter ε which characterises some rapid variation in the original problem. One then treats the scaled and unscaled parameters as independent variables and introduces a formal asymptotic expansion in ascending powers of ε, which in turn generates a hierarchy of cell problems. These cell problems may be solved, analytically or otherwise, to obtain effective coefficients or material parameters. Several extensive monographs [10, 12, 172] have been written on the topic of asymptotic homogenisation, primarily focusing on statics, low, and high frequencies.

Here we briefly review the approach of Craster et al. outlined in [47, 55]. It is similar to classical asymptotic homogenisation in the sense that it makes use of scaled variables; but, rather than perturbing away from a static configuration, the method of Craster et al. involves perturbations near resonances and results in a hierarchy of model eigenvalue problems. We remark that asymptotic homogenisation of spectral problems was also treated by Papanicolaou [12] and, more recently, by many others; the reader is referred to the excellent review article [180], and references therein, for more detail.

2.7.1 Returning to the biatomic chain

We now return to the biatomic chain considered in Sections 2.2 and 2.6 and assume time-harmonic motion of radian frequency ω such that the homogeneous equations of motion are

$$-m_i\omega^2 u_{i,n} = u_{j,n} + u_{j,n+(-1)^i} - 2u_{i,n}, \quad i,j = 1,2,\ i \neq j, \quad (2.85)$$

where summation over repeated indicies is not assumed. The indices $i,j = 1,2$ enumerate the two different masses in the same elementary cell, whilst n identifies the elementary cell itself. The left-hand side of (2.85) is the dynamic term involving the inertia of the two masses m_1 and m_2; the right-hand side are the spring reaction forces resulting from the displacements of the two neighbouring masses, as well as the ith mass itself.

We begin by introducing a small parameter $0 < \varepsilon \ll 1$, which characterises the ratio between two disparate length scales. In particular, since the bond length is unitary, we may define $\varepsilon = 1/N$ with $N \gg 1$ being the number of lattice points in the biatomic chain. With this small parameter in place, we formally introduce two scales: the first is long-scale or slow-scale $\xi = \varepsilon n$, which we treat as a continuous variable; the second length scale is the short-scale or fast-scale which is characterised by the discrete variable $p \in \{-1, 0, 1\}$. These two variables are considered as being independent and we formally set $u_{i,n+p} = v_i(\xi + \varepsilon p, p)$ and introduce the vector $\boldsymbol{v}(\xi,p) = (v_1(\xi,p), v_2(\xi,p))^{\text{T}}$. Since ξ is continuous, the displacement admits the representation

$$\boldsymbol{v}(\xi + \varepsilon p, p) \sim \boldsymbol{v}(\xi,p) + \varepsilon p \frac{\partial}{\partial \xi}\boldsymbol{v}(\xi,p) + \varepsilon^2 \frac{p^2}{2} \frac{\partial^2}{\partial \xi^2}\boldsymbol{v}(\xi,p), \quad \text{for} \quad 0 < \varepsilon \ll 1,$$

and Equation (2.85) becomes

$$0 = u_j(\xi,0) + \left(m_i\omega^2 - 2\right)u_i(\xi,0) + u_j(\xi,(-1)^i)$$
$$+ (-1)^i \varepsilon u_j'(\xi,(-1)^i) + \varepsilon^2 \frac{1}{2} u_j''(\xi,(-1)^i) + \mathcal{O}(\varepsilon^3), \quad (2.86)$$

where the superscript primes denote successive differentiation with respect to ξ.

Fundamentally the dynamic homogenisation scheme of Craster et al. [55], as well as other similar schemes for the asymptotic analysis of spectral problems (see, for example, the work by Papanicolaou [12]), involves perturbations away from standing waves. To this end we impose the condition

$$\boldsymbol{v}(\xi, \pm 1) = (-1)^s \boldsymbol{v}(\xi, 0), \quad s = 0, 1. \quad (2.87)$$

The condition (2.87) relates the displacements of the nodes in cell n to those of the neighbouring masses on the short-scale; physically, this condition corresponds to in-phase standing waves when $s = 0$ and out-of-phase resonances when $s = 1$. It is important to note that this resonance condition is applied only on the short-scale.

Foundations and methods

Imposing the short-scale resonance condition (2.87) on the displacements in (2.86) yields

$$0 = \left\{ \begin{pmatrix} m_1\omega^2 - 2 & 1+(-1)^s \\ 1+(-1)^s & m_2\omega^2 - 2 \end{pmatrix} + \varepsilon \begin{pmatrix} 0 & (-1)^{s+1}\partial \\ (-1)^s\partial & 0 \end{pmatrix} \right.$$
$$\left. + \frac{\varepsilon^2}{2} \begin{pmatrix} 0 & (-1)^s\partial^2 \\ (-1)^s\partial^2 & 0 \end{pmatrix} \right\} v(\xi), \quad (2.88)$$

where $\partial = \partial/\partial\xi$ and we have suppressed the second argument of $v(\xi,0)$. We note that the leading order matrix in (2.88) is diagonal for $s = 1$, which implies that the motion of the two masses in the chain de-couple at leading order. The form of (2.88), in ascending powers of ε, suggests the following ansatz for the displacement and frequency-squared

$$v(\xi) = \sum_{q=0}^{\infty} \varepsilon^q V_q(\xi), \qquad \omega^2 = \sum_{q=0}^{\infty} \varepsilon^q \omega_q^2. \quad (2.89)$$

Substitution of the ansatz into (2.88) provides a hierarchy of equations in powers of ε, the first three of which are,

$$\begin{pmatrix} m_1\omega_0^2 - 2 & 1+(-1)^s \\ 1+(-1)^s & m_2\omega_0^2 - 2 \end{pmatrix} V_0(\xi) = 0, \quad (2.90)$$

$$\begin{pmatrix} m_1\omega_0^2 - 2 & 1+(-1)^s \\ 1+(-1)^s & m_2\omega^2 - 2 \end{pmatrix} V_1(\xi) = \begin{pmatrix} -m_1\omega_1^2 & (-1)^s\partial \\ (-1)^{s+1}\partial & -m_2\omega_1^2 \end{pmatrix} V_0(\xi), \quad (2.91)$$

$$\begin{pmatrix} m_1\omega_0^2 - 2 & 1+(-1)^s \\ 1+(-1)^s & m_2\omega_0^2 - 2 \end{pmatrix} V_2(\xi) = \begin{pmatrix} -m_1\omega_1^2 & (-1)^s\partial \\ (-1)^{s+1}\partial & -m_2\omega_1^2 \end{pmatrix} V_1(\xi)$$
$$+ \frac{1}{2}\begin{pmatrix} -m_1\omega_2^2 & (-1)^{s+1}\partial^2 \\ (-1)^{s+1}\partial^2 & -m_2\omega_2^2 \end{pmatrix} V_0(\xi). \quad (2.92)$$

2.7.1.1 Leading order problem

The leading order problem (2.90) is a standard eigenvalue problem and, for $m_1 \neq m_2$, has two distinct eigenvalues given by

$$\omega_0^2 = \frac{m_1 + m_2}{m_1 m_2} \pm \sqrt{\left(\frac{m_1+m_2}{m_1 m_2}\right)^2 - 2\frac{1+(-1)^{s+1}}{m_1 m_2}}. \quad (2.93)$$

We note that this equation is equivalent to the dispersion equation (2.71) derived earlier for the two-mass chain, if we set $k = 0$ and $k = \pi$ in (2.71), i.e. at the resonant points. The frequencies of the in-phase standing waves ($s = 0$) are then

$$\omega_0^{(1)} = 0, \qquad \omega_0^{(2)} = \sqrt{\frac{2}{m_1} + \frac{2}{m_2}},$$

and for the out-of-phase standing waves ($s = 1$)

$$\omega_0^{(3)} = \sqrt{\frac{2}{m_1}}, \qquad \omega_0^{(4)} = \sqrt{\frac{2}{m_2}}.$$

The parenthesised superscript number index the modes and we shall use this enumeration to refer back to these particular modes.

It is clear that the leading order problem (2.90) does not explicitly depend on the long-scale variable ξ and, therefore, the solution permits the decomposition $\boldsymbol{V}_0(\xi) = f_0(\xi)\boldsymbol{\mathcal{V}}_0$, where $\boldsymbol{\mathcal{V}}_0$ is the eigenvector of the matrix appearing in (2.90) and $f_0(\xi)$ is an arbitrary modulating function. In this case, the corresponding eigenvectors are

Mode 1: $\qquad \boldsymbol{\mathcal{V}}_0^{(1)} = (1, 1)^{\mathrm{T}}, \qquad\qquad \omega_0^{(1)} = 0,$

Mode 2: $\qquad \boldsymbol{\mathcal{V}}_0^{(2)} = (m_2, -m_1)^{\mathrm{T}}, \qquad \omega_0^{(2)} = \sqrt{\frac{2}{m_1} + \frac{2}{m_2}},$

Mode 3: $\qquad \boldsymbol{\mathcal{V}}_0^{(3)} = (1, 0)^{\mathrm{T}}, \qquad\qquad \omega_0^{(3)} = \sqrt{\frac{2}{m_1}},$

Mode 4: $\qquad \boldsymbol{\mathcal{V}}_0^{(4)} = (0, 1)^{\mathrm{T}}, \qquad\qquad \omega_0^{(4)} = \sqrt{\frac{2}{m_2}}.$

In order to complete the leading order solution, it now remains to determine the corresponding modulating functions $f_0^{(i)}(\xi)$.

2.7.1.2 Next-to-leading order problem

Proceeding to the next order, it is clear from the previous discussion in Section 2.7.1.1 that the matrix on the left-hand side of (2.91) is degenerate. Thus, in order to guarantee the existence of a solution, we require that [159]

$$\boldsymbol{\mathcal{V}}_0^{(i)\dagger} \begin{pmatrix} -m_1\omega_1^2 & (-1)^s\partial \\ (-1)^{s+1}\partial & -m_2\omega_1^2 \end{pmatrix} \boldsymbol{\mathcal{V}}_0^{(i)} f_0^{(i)}(\xi) = 0, \qquad (2.94)$$

where $(\cdot)^\dagger$ denotes the Hermitian transpose. Usually, and for all the cases considered here, the left-hand side of (2.94) vanishes if and only if $\omega_1 \equiv 0$ in

Foundations and methods

which case, the next-to-leading order solution is of the form

$$\begin{aligned}\boldsymbol{V}_1^{(i)} &= \begin{pmatrix} m_1\omega_0^2 - 2 & 1+(-1)^s \\ 1+(-1)^s & m_2\omega_0^2 - 2 \end{pmatrix}^+ \begin{pmatrix} 0 & (-1)^s\partial \\ (-1)^{s+1}\partial & 0 \end{pmatrix} \boldsymbol{V}_0^{(i)} f_0^{(i)}(\xi) \\ &\quad - \left\{ \begin{pmatrix} m_1\omega_0^2 - 2 & 1+(-1)^s \\ 1+(-1)^s & m_2\omega_0^2 - 2 \end{pmatrix}^+ \begin{pmatrix} m_1\omega_0^2 - 2 & 1+(-1)^s \\ 1+(-1)^s & m_2\omega_0^2 - 2 \end{pmatrix} \right. \\ &\quad \left. - \begin{pmatrix} 1 & 0 \\ 0 & 1 \end{pmatrix} \right\} \boldsymbol{A}(\xi), \quad (2.95)\end{aligned}$$

where $(\cdot)^+$ denotes the Moore–Penrose pseudoinverse and $\boldsymbol{A}(\xi)$ is an arbitrary vector function, which is not required for the remaining discussion (also see [47]).

2.7.1.3 Second-order problem

Moving to the second-order problem, we find that the solvability condition yields

$$\boldsymbol{V}_0^{(i)\dagger} \left(\mathcal{D}_1^{(s)} + \mathcal{D}_2^{(s)}\right) \boldsymbol{V}_0^{(i)} f_0^{(i)}(\xi) = 0, \quad (2.96)$$

where

$$\mathcal{D}_1^{(s)} = \begin{pmatrix} 0 & (-1)^s\partial \\ (-1)^{s+1}\partial & 0 \end{pmatrix} \begin{pmatrix} m_1\omega_0^2 - 2 & 1+(-1)^s \\ 1+(-1)^s & m_2\omega_0^2 - 2 \end{pmatrix}^+ \begin{pmatrix} 0 & (-1)^s\partial \\ (-1)^{s+1}\partial & 0 \end{pmatrix},$$

$$\mathcal{D}_2^{(s)} = \frac{1}{2}\begin{pmatrix} -2m_1\omega_2^2 & (-1)^{s+1}\partial^2 \\ (-1)^{s+1}\partial^2 & -2m_2\omega_2^2 \end{pmatrix}.$$

Provided that $\omega_2 \neq 0$, Equation (2.96) is a second-order ordinary differential equation that governs the long-scale behaviour of the system; we note that (2.96) involves only the long-scale variable ξ and second-order correction ω_2 to the frequency.

Mode 1. For $s = 0$ and $i = 1$, the second-order solvability condition (2.96) yields the homogenised equation for the long-scale behaviour

$$\left[\frac{d^2}{d\xi^2} + 2(m_1 + m_2)\omega_2^2\right] f_0^{(1)}(\xi) = 0. \quad (2.97)$$

This corresponds to the acoustic branch near the origin, i.e. in the quasi-static regime where classical asymptotic homogenisation may be used. The homogenised differential equation (2.97) is posed entirely on the long-scale and indicates that the two-mass chain behaves like a continuous string or rod. On the short-scale, the eigenvector is $\boldsymbol{V}_0^{(1)} = (1,1)^T$, which means the chain undergoes rigid-body motion on the microscale, with the two masses moving in-phase.

The asymptotic dispersion equation in the vicinity of $\omega = 0, k = 0$, can be obtained by formally searching for solutions to the homogenised equation (2.97) of the form $f_0^{(1)} = e^{ik\xi/\varepsilon}$, i.e. Bloch modes. In this case, we have

$$\omega^2 \sim \frac{k^2}{2(m_1 + m_2)}, \qquad (2.98)$$

which agrees with the asymptotic expansion of the exact dispersion equation (2.71) in the limit $k \to 0$.

Mode 2. This mode, $s = 0$ and $i = 2$, corresponds to the in-phase ($k = 0$) resonance of the optical mode. Once again, we find that $\omega_1 = 0$ and the solvability condition (2.96) yields a homogenised equation very similar to (2.98) for the acoustic band

$$\left[\frac{d^2}{d\xi^2} - 2(m_1 + m_2)\omega_2^2\right] f_0^{(2)}(\xi) = 0. \qquad (2.99)$$

As previously, the asymptotic dispersion equation can be obtained by setting $f_0^{(2)} = e^{ik\xi/\varepsilon}$ and, in this case, we obtain

$$\omega^2 \sim 2\left(\frac{1}{m_1} + \frac{1}{m_2}\right) - \frac{k^2}{2(m_1 + m_2)}. \qquad (2.100)$$

We note the opposing signs in Equations (2.97) and (2.99); this difference manifests itself in the behaviour on the long-scale for perturbations around the resonance. In particular, the operators in (2.97) and (2.99) admit eigenfunctions of the form

$$f_0^{(1)} = \exp\left(i|\xi|\omega_2\sqrt{2(m_1 + m_2)}\right), \qquad (2.101a)$$

and

$$f_0^{(2)} = \exp\left(-|\xi|\omega_2\sqrt{2(m_1 + m_2)}\right). \qquad (2.101b)$$

Clearly, the behaviour of the eigensolutions depends on the value of ω_2. If $\omega > \omega_0$, corresponding to the real ω_2, then the long-scale solutions near $\omega = 0$ will be oscillatory whilst the solution near the optical resonance will be decaying. On the other hand, if $\omega < \omega_0$ such that ω_2 is purely imaginary, then the solutions near the optical resonance will become oscillatory. This behaviour can be interpreted in terms of perturbations, linked to the lattice Green's function, as discussed in Section 2.6.

The eigenvector in this case is $\mathbf{V}_0^{(2)} = (m_2, -m_1)^T$, which corresponds to the two disparate masses oscillating out-of-phase, whilst all nodes with the same mass oscillate in-phase.

Mode 3. Moving to the opposite end of the irreducible Brillouin zone, near $k = \pi$, we set $s = 1$ and choose $i = 3$. In this case, ω_1 once again vanishes, and (2.96) reduces to

$$\left[\frac{d^2}{d\xi^2} - (m_1 - m_2)\omega_2^2\right] f_0^{(3)}(\xi) = 0, \qquad (2.102)$$

where, without loss of generality, we assume that $m_1 > m_2$ as in Section 2.6. Hence, on the long-scale, the modulating function is

$$f_0^{(3)} = \exp\left(-|\xi|\omega_2\sqrt{2(m_1 - m_2)}\right), \tag{2.103}$$

which is oscillatory for $\omega < \omega_0^{(3)}$ and exponentially decaying for $\omega > \omega_0^{(3)}$. The corresponding asymptotic dispersion equation is

$$\omega^2 \sim \frac{2}{m_1} - \frac{(\pi - k)^2}{2(m_1 - m_2)}. \tag{2.104}$$

The eigenvector in this case is $\boldsymbol{V}_0^{(3)} = (1, 0)^{\mathrm{T}}$, which means that the heavier masses oscillate out-of-phase with each other whilst the lighter masses remain stationary.

Mode 4. We now arrive at the final mode with $s = 1$ and $i = 4$. Here, ω_1 once again vanishes, and (2.96) reduces to

$$\left[\frac{\mathrm{d}^2}{\mathrm{d}\xi^2} + 2(m_1 - m_2)\omega_2^2\right] f_0^{(4)}(\xi) = 0, \tag{2.105}$$

In contrast to mode 3, the modulating function

$$f_0^{(4)} = \exp\left(i|\xi|\omega_2\sqrt{2(m_1 - m_2)}\right), \tag{2.106}$$

is oscillatory for $\omega > \omega_0^{(4)}$ and exponentially decaying for $\omega < \omega_0^{(4)}$. The corresponding asymptotic dispersion equation is

$$\omega^2 \sim \frac{2}{m_2} + \frac{(\pi - k)^2}{2(m_1 - m_2)}. \tag{2.107}$$

The eigenvector is $\boldsymbol{V}_0^{(4)} = (0, 1)^{\mathrm{T}}$, which means that the lighter masses oscillate out-of-phase with each other whilst the heavier masses remain stationary.

2.7.2 Propagation and decay

Figure 2.8 shows the dispersion curves for the two-mass chain (with $m_1 = 1$ and $m_2 = 2$). The solid curves are the exact solutions to the dispersion equation, whilst the dashed lines are their asymptotic approximations in the neighbourhoods of the associated resonances. The exact dispersion curves and associated spectrum have already been discussed in Section 2.6.

The modulating functions (2.101a), (2.101b), (2.103), and (2.106) prove convenient when analysing the macroscopic behaviour of the two-mass chain in the vicinity of resonances. For instance, consider the semi-infinite band gap $2/m_2 < \omega^2 < 2/m_1$. The band gap is bounded from above by the resonance associated with mode 4 and from below by that of mode 3; these modes

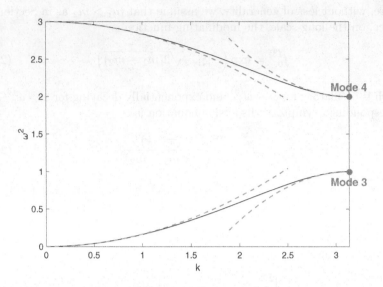

FIGURE 2.8: The dispersion curves for the biatomic chain. The solid black curves are the exact solutions to the dispersion equation, whilst the dashed grey lines are the next-to-leading order asymptotic curves in the vicinity of the relevant resonances.

are near $k = \pi$ and, hence, are anti-symmetric. As noted above, the two modulating functions, (2.103) and (2.106), have contrasting behaviour but are related in such a way that $f_0^{(3)}(-i\omega_2) = f_0^{(4)}(\omega_2)$. This means that if we perturb away from the resonances where $\omega \sim \omega_0^2 + \varepsilon^2 \omega_2^2$, then the two modulating functions will have contrasting behaviour: one will decay as $n \to \infty$, whilst the other will be oscillatory.

This effect is illustrated in Table 2.1, where we plot the leading order behaviour of the two-mass chain near the two resonant modes identified on the dispersion diagram in Figure 2.8. In particular, if we perturb the frequency *downwards* such that $\omega < \omega_0$ and ω_2 is purely imaginary, then we obtain the behaviour illustrated in the first row of Table 2.1. Perturbing *downwards* from the resonance associated with the upper branch of the dispersion curves pushes mode four into the finite band gap; whereas mode three is pushed into the pass band, hence the respective decay and propagation. The converse is true for *upward* perturbations with $\omega > \omega_0$, as shown on the second row of Table 2.1.

2.7.2.1 Comparison with the exact approach

The exact solution for modes inside the finite band gap was obtained in Section 2.6.2, in the form of the stop-band Green's function. In particular, the

TABLE 2.1: The leading order, $V_0(\xi)$, behaviour of the two-mass chain in the neighbourhood of the two band-edge modes indicated in Figure 2.8.

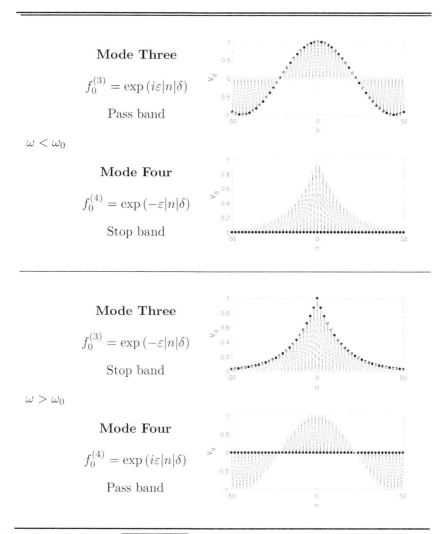

$\omega < \omega_0$

Mode Three

$f_0^{(3)} = \exp\left(i\varepsilon |n|\delta\right)$

Pass band

Mode Four

$f_0^{(4)} = \exp\left(-\varepsilon |n|\delta\right)$

Stop band

$\omega > \omega_0$

Mode Three

$f_0^{(3)} = \exp\left(-\varepsilon |n|\delta\right)$

Stop band

Mode Four

$f_0^{(4)} = \exp\left(i\varepsilon |n|\delta\right)$

Pass band

Note: Here $\delta = |\omega_2 \sqrt{2(m_1 - m_2)}|$. The black and grey dots indicate the nodes with mass $m_1 = 2$ and $m_2 = 1$, respectively. The solid black lines indicate the interatomic bonds and the dashed grey curves indicate the modulating functions $f_0^{(i)}(\xi)$. The numerical values used were $\varepsilon = 0.1$ and $|\omega_2| = 0.5$.

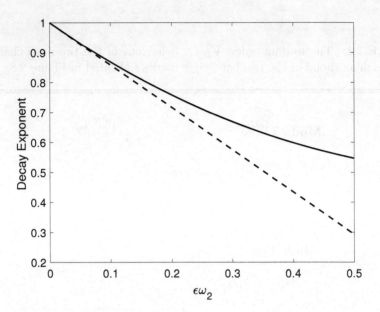

FIGURE 2.9: A comparison for the decay rates obtained using the exact Green's function approach, $|\lambda|$ (solid line), and the asymptotic homogenisation approach, $|\Lambda| = |1 - \varepsilon\omega_2\sqrt{2(m_1 - m_2)}|$ (dashed line).

solutions take the form

$$u_{1,n} = C_{1,0}\lambda^{|n|} + C_{1,1}\lambda^{|n-1|}, \qquad u_{2,n} = C_{2,0}\lambda^{|n|} + C_{2,1}\lambda^{|n+1|}, \qquad (2.108)$$

where $\lambda = (\sqrt{Q_0^2 - 4Q_0} + Q_0 - 2)/2$ is the *localisation exponent* which characterises the decay rate and $Q_0 = (m_1\Omega^2 - 2)(m_2\Omega^2 - 2) < 0$ for $2/m_1 < \omega^2 < 2/m_2$, i.e. in the stop band. Consider a perturbation into the stop band from the lower boundary of the finite band gap (mode three) such that $\omega^2 \sim \omega_0^2 + \varepsilon^2\omega_2^2$ with $\omega_2 \in \mathbb{R}$ and $0 < \varepsilon \ll 1$, whence

$$\lambda \sim -1 + \varepsilon\omega_2\sqrt{2(m_1 - m_2)}. \qquad (2.109)$$

The corresponding modulating function obtained through asymptotic homogenisation is (2.103)

$$f_0^{(3)} = \Lambda^{|n|}, \qquad (2.110)$$

where we have introduced an analogous *localisation exponent* $\Lambda = \exp\left(-\varepsilon\omega_2\sqrt{2(m_1 - m_2)}\right)$. Expanding this *localisation exponent* for small ε,

$$\Lambda \sim 1 - \varepsilon\omega_2\sqrt{2(m_1 - m_2)}, \qquad (2.111)$$

it is easy to observe that the decay rates $|\lambda| = |\Lambda|$ agree in the asymptotic limit $\varepsilon \to 0$.

Similarly for perturbations from the upper boundary of the finite band gap, such that $\omega^2 \sim \omega_0^2 - \varepsilon^2 \omega_2^2$ with $\omega_2 \in \mathbb{R}$, we find

$$\lambda \sim -1 + \varepsilon\omega_2\sqrt{2(m_1 - m_2)}, \quad \text{and} \quad \Lambda \sim 1 - \varepsilon\omega_2\sqrt{2(m_1 - m_2)}, \quad (2.112)$$

and $|\lambda| = |\Lambda|$.

Although we have focused on a relatively simple case, the asymptotic theory developed by Craster and co-workers and summarised in the present section can be applied to a wide range of continuous and discrete periodic structures. The method of dynamic homogenisation is attractive for several reasons. In particular, it allows for straightforward analysis of the dynamic response of structured media in the vicinity of any resonance, not simply in the quasi-static regime where classical asymptotic homogenisation is valid. Moreover, as we have just seen, the method of Craster et al. agrees well with the exact solution in the neighbourhood of resonances.

Having reviewed the necessary fundamental concepts and approaches, in the next chapter we will examine the propagation of waves in multi-scale media composed of thin ligaments.

Chapter 3

Waves in structured media with thin ligaments and disintegrating junctions

Vibration absorption is a well-established technique in the design, analysis and engineering of seismic-resistant structures. For the propagation of elastic waves in composite structures and periodic media, many additional issues are raised, compared to wave propagation in homogeneous structures, which include, for example, effects of filtering and polarisation.

For structures exhibiting periodicity, it will be shown that small resonators, placed within each of the elementary cells, may be used to control the propagation of elastic waves through the entire structure. The properties of such resonators may be altered in order to tune them to the desired behaviour of wave propagation. Such resonators may have the advantage of being small in comparison with the size of the overall periodic structure or wavelength of the wave propagating through the periodic structure. These are termed "multi-scale" resonators. Structures may be designed to include multi-scale resonators to achieve a desired dynamic response and wave propagation properties.

In this chapter, structures containing multi-scale resonators will be analysed, with the emphasis on a class of standing waves of sufficiently low frequencies. The propagation of waves through finite structured solids will be related to the propagation of Bloch–Floquet waves in the corresponding infinite periodic structures. It will be shown how these can be used in the modelling of wave propagation across structured interfaces, localisation and negative refraction.

Defect formation in the resonators embedded into the multi-scale structure will be studied, and a comparison will be made between structures containing damaged and undamaged resonators. This chapter is based on the results of the papers [69, 82, 83].

3.1 Structures with undamaged multi-scale resonators

In [130, 133, 166] explicit, closed form solutions of the time-harmonic wave problems were derived via the multipole method, for the case of arrays of circular inclusions.

It is possible to design structures, possessing stop bands, by incorporating appropriately designed resonators within the elementary cell of the (periodic) structure (see, for example, [126]). In particular, when a multi-scale structure has components of high contrast, filtering of waves of a sufficiently large length can be achieved even in the case of elementary cells being small in size. The applications may range from acoustic filters to filters for elastic waves in earthquake-resistant structures. Here, we also discuss a possible mechanism to control the stiffness of resonators within an elementary cell and hence to alter the position of the stop bands on the dispersion diagram. In this configuration, the control parameter is temperature.

For the full vector problem of planar elasticity, a design of structure is presented which filters waves of low frequency. A mechanism to control the frequency of the stop bands is also described, using temperature as the control parameter. The thermally inhomogeneous structure includes elements that buckle under the thermal load when combined with the mechanical constraints imposed by the overall structure. The spectral characteristics of the modified structure are thus changed and this is reflected in the dispersion diagram.

The governing equations and geometry of the periodic structure are described in Section 3.1.1. The mechanism of the thermal pre-stress is given in Section 3.1.2 and an evaluation of the critical temperature required to buckle a thin ligament within the structure is also presented. Asymptotic estimates of the frequencies of standing waves, of rotational and translational types, and existing at the boundaries of the stop bands of the dispersion diagrams, are given in Section 3.1.3. Finally, the results of numerical simulations and comparison of asymptotic estimates and numerical computations are presented in Section 3.1.4.

3.1.1 Geometry and governing equations

Bloch–Floquet waves in an infinite periodic two-dimensional elastic structure, whose elementary cell is shown in Figure 3.1, are studied.

The cell incorporates a resonator consisting of a relatively large solid Ω_1 and thin rods Π_j, $j = 1, \ldots, N$, of equal length l, with N being a given positive integer. The thin rods are attached to the surrounding frame Ω_2. The cell of periodicity is denoted by $\Omega = \Omega_1 \cup \Omega_2 \cup_{j=1}^{N} \Pi_j$. The interior boundary within the elementary cell is denoted by Γ, whereas γ represents the exterior boundary of the cell. The elastic displacement is a time-harmonic vector function, whose amplitude $\mathbf{u}(\mathbf{x})$ satisfies the following equations:

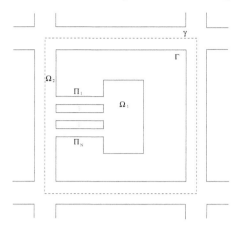

FIGURE 3.1: The elementary cell of the multi-scale periodic structure.

$$\mu \Delta \mathbf{u}(\mathbf{x}) + (\lambda + \mu)\nabla\nabla \cdot \mathbf{u}(\mathbf{x}) + \rho\omega^2 \mathbf{u}(\mathbf{x}) = \mathbf{0}, \quad \mathbf{x} = (x_1, x_2) \in \Omega, \qquad (3.1)$$

$$\boldsymbol{\sigma}^{(n)}(\mathbf{u}; \mathbf{x}) = \mathbf{0}, \quad \mathbf{x} \in \Gamma, \qquad (3.2)$$

where $\mathbf{u} = (u_1, u_2)$, and $\boldsymbol{\sigma}^{(n)}$ is the vector of tractions with the entries $\sigma_j^{(n)} = \sum_k \sigma_{jk} n_k$, with σ_{jk} being the components of stress and $\mathbf{n} = (n_1, n_2)^T$ being the outward unit normal on the interior boundary Γ. Here λ and μ are the Lamé elastic moduli, ρ is the mass density and ω is the radian frequency.

The Bloch–Floquet quasi-periodicity conditions are also imposed on the vector function $\mathbf{u}(\mathbf{x})$ as follows:

$$\mathbf{u}(\mathbf{x} + m_1 \mathbf{a}_1 + m_2 \mathbf{a}_2) = \mathbf{u}(\mathbf{x}) e^{i(K_1 m_1 + K_2 m_2)}, \qquad (3.3)$$

for all $\mathbf{x} \in \Omega$, and integer m_1, m_2,

where $\mathbf{a}_1, \mathbf{a}_2$ are the basis vectors of the doubly periodic structure and $\mathbf{K} = (K_1, K_2)^T$ is the Bloch vector.

It is assumed that the above structure is heated in such a way that the temperature rises uniformly to a constant, time independent value. In this case, the spatial derivatives of the temperature are zero, and hence the temperature does not appear in the equations of motion. However, it is also assumed that the thermal expansion coefficients are different for the thin rods and for the remaining part of the elementary cell: α stands for the thermal expansion coefficient of the materials occupying Ω_1 and Ω_2, whereas α_j correspond to the thermal expansion coefficients of thin rods Π_j.

The aim is to study the solutions of the spectral problem (3.1)–(3.3), with particular emphasis on the dispersion diagrams and potential stop bands. The

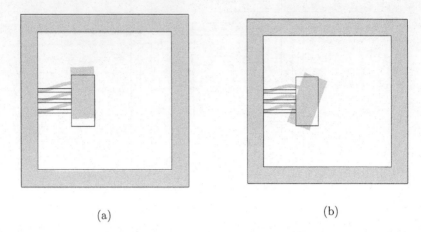

(a) (b)

FIGURE 3.2: Standing waves within the periodic structure; (a) the low-frequency translational eigenmode, (b) the higher frequency rotational eigenmode.

particular feature of the configuration shown in Figure 3.1 is that the thin rods may be subjected to Euler's buckling due to the thermally induced pre-stress. In this case, the stiffness of the overall system changes and this leads to changes in the dispersion diagram and, in particular, the position of any stop bands. The analytical model and the corresponding numerical simulations are presented below. For the purpose of illustration, we refer to the structure within the frame in Figure 3.2 with three thin elastic rods, where the middle rod has a relatively large thermal expansion coefficient. Consequently, this leads to the middle rod buckling and hence a reduction of the stiffness of the elastic structure when a sufficiently high temperature is applied.

3.1.2 Thermal pre-stress and Euler's buckling

Temperature is used here as a control parameter. A pre-stress is induced which alters the stiffness of the structure. It is assumed that the structure Ω is heated uniformly, with an increase in temperature of δT. Consider the case of the structure with three rods, shown in Figure 3.2. It is assumed that all the material parameters of the thin rods are the same except for the thermal expansion coefficients. Namely, the thermal expansion coefficient α_2 for the middle rod is larger than that of other rods α_1. For a certain critical temperature, the middle rod, subjected to compressive pre-stress due to the thermal load, will buckle.

Treating Ω_1 as a rigid solid and assuming that the extension of all thin rods is the same, we derive the expression for the compressive force within Π_2 induced by the change in temperature δT:

$$P_2 = \frac{2\delta T(\alpha_2 - \alpha_1)Eh}{3}, \tag{3.4}$$

where E is the Young's modulus, and h is the width of the rod.

The critical load for the Euler's buckling of the rod Π_2 is given by (see [182])

$$P_{cr} = \frac{4\pi^2 Eh^3}{12l^2}. \tag{3.5}$$

Hence, the formulae (3.4) and (3.5) yield the expression for the critical temperature for buckling δT_{cr} as

$$\delta T_{cr} = \frac{1}{2}(\alpha_2 - \alpha_1)^{-1}\pi^2 (h/l)^2. \tag{3.6}$$

It is noted that h/l is a small parameter and hence δT_{cr} becomes smaller for thinner rods of the same length and fixed positive $\alpha_2 - \alpha_1$.

As soon as the critical value of the temperature is reached, it is assumed that the contribution from the buckled rod, as a stiffness element, can be neglected. Correspondingly, the structure with three rods connecting the body Ω_1 to the surrounding matrix is replaced by a structure of reduced stiffness, with the two outer rods only. The eigenfrequency corresponding to a particular vibration mode, shown in Figure 3.2a, will become smaller when the number of thin rods, contributing as stiffness elements, is reduced as a result of this thermal buckling. This is discussed further in the next section.

3.1.3 Asymptotic approximations for two standing wave modes

The frequencies of some standing waves can be approximated analytically. In particular, for the case of the multi-scale structure with built-in resonators we distinguish between the modes representing translational and rotational motion and derive the approximation formulae for the corresponding frequencies.

3.1.3.1 Fundamental translational mode

Figure 3.2a presents an eigenmode corresponding to a standing wave within the periodic structure, such that the subdomain Ω_1 of the resonator moves in the vertical direction, with a low frequency. In this section, the analytical asymptotic approximation for the corresponding eigenvalue is derived. For this particular case, the displacement components satisfy the periodicity conditions on the exterior boundary of the elementary cell. This is equivalent to setting the components K_1, K_2 of the Bloch vector to be zero in (3.3).

For the periodic case and non-zero frequency, the average of $\rho\omega^2 u_j$ over the elementary cell Ω is equal to zero.

It is observed for this vibration mode (see Figure 3.2a) that the subdomains Ω_1 and Ω_2 move like rigid solids in the vertical direction, perpendicular to the thin rods. Let $C_j \mathbf{e}^{(2)}$ be the displacements of the bodies Ω_j, respectively. Neglecting the inertia of thin rods, this leads to

$$\rho_1 C_1 |\Omega_1| + \rho_2 C_2 |\Omega_2| = 0,$$

where $|\Omega_j|$, $j = 1, 2$, are areas of Ω_j.

Consider one of the thin rods, say Π_k, and let $u_2^{(k)}(x_1)$ be the transverse displacement of the points of this rod. Neglecting the inertia of the rod, then the function $u_2^{(k)}$ is the solution of the boundary value problem

$$(u_2^{(k)})''''(x_1) = 0, \quad x_1 \in (0, l),$$

$$u_2^{(k)}(0) = C_2 = -C_1 |\Omega_1|/|\Omega_2|, \quad (u_2^{(k)})'(0) = 0,$$

$$u_2^{(k)}(l) = C_1, \quad (u_2^{(k)})'(l) = 0.$$

Hence, $u_2^{(k)}$ is a cubic function

$$u_2^{(k)}(x_1) = A x_1^3 + B x_1^2 + D,$$

where

$$A = -2C_1 \frac{1 + |\Omega_1|/|\Omega_2|}{l^3}, \quad B = 3C_1 \frac{1 + |\Omega_1|/|\Omega_2|}{l^2}, \quad D = -C_1 \frac{|\Omega_1|}{|\Omega_2|}.$$

The stiffness coefficient s_k for the rod Π_k is defined by

$$s_k = E h_k^3 / (12(1 - \nu^2)),$$

with E, ν being Young's modulus and Poisson's ratio, respectively; h_k stands for the thickness of Π_k.

The transverse force exerted by Π_k on the body Ω_1 is equal to

$$\mathcal{F}^{(k)} = s_k (u_2^{(k)})'''(l) = -12 s_k C_1 (1 + |\Omega_1|/|\Omega_2|)/l^3.$$

Taking into account contributions from all the rods and writing the equation of motion for Ω_1, then

$$\rho |\Omega_1| \omega_t^2 C_1 = 12 C_1 (1 + |\Omega_1|/|\Omega_2|) \sum_{j=1}^{N} s_j / l^3.$$

Hence the corresponding (translational) eigenfrequency ω_t is approximated by

$$\omega_t \approx \left(\frac{(1 + |\Omega_1|/|\Omega_2|) E}{\rho |\Omega_1| (1 - \nu^2) l^3} \sum_{j=1}^{N} h_j^3 \right)^{1/2}. \tag{3.7}$$

3.1.3.2 Fundamental rotational mode

Another interesting vibration mode corresponds to a standing wave and is shown in Figure 3.2b. This is a rotational mode, and an asymptotic estimate will be provided below for the corresponding radian frequency ω_r.

Let Ω_1 be a rectangular block $\Omega_1 = \{(x_1, x_2) : l < x_1 < l + a, |x_2| < b/2\}$, and let the distance between the rods Π_1 and Π_3 be $2d$. We consider the two-rod structure, since the contribution of the middle rod Π_2 to the frequency of this rotational mode is negligibly small, as it lies on the axis of symmetry of the resonator.

It is assumed in this asymptotic approximation that Ω_1 is a rigid body, and the corresponding equation of balance of angular momentum yields

$$\mathbf{r} \times \mathbf{F}^{(1)} + \mathbf{r} \times \mathbf{F}^{(3)} = -\omega_r^2 A I_{\Omega_1} \mathbf{e}^{(3)}, \tag{3.8}$$

where $\mathbf{F}^{(j)}$ are the force vectors exerted on the body Ω_1 by the thin rods Π_j, A is the angle of rotation, ω_r is the rotational eigenfrequency, and I_{Ω_1} is the moment of inertia of Ω_1.

Assuming that the inertia of the thin rods is negligibly small and hence approximating the longitudinal displacements as linear functions in x_1 we obtain that

$$\mathbf{F}^{(1)} = -\mathbf{F}^{(3)} \approx EhAd/l\ \mathbf{e}^{(1)}. \tag{3.9}$$

Here the small transverse component of the force is neglected. Then Equation (3.8) can be approximated in the form

$$2Ehd^2/l - \omega_r^2 I_{\Omega_1} \approx 0,$$

which yields the approximation for the frequency of the rotational vibration mode shown in Figure 3.2b:

$$\omega_r \approx \left(\frac{2Ehd^2}{lI_{\Omega_1}}\right)^{1/2}. \tag{3.10}$$

3.1.4 Dispersion diagrams and stop bands

Here, we discuss three sets of dispersion curves for Bloch–Floquet waves, the first of which is in the doubly periodic structure shown in Figure 3.1 but without the three-legged resonator. This will be referred to as the frame structure. The second structure includes the three-legged resonator as shown in Figure 3.1. The third structure is similar to the second but with the middle leg of the resonator being removed. This will be referred to as the "two-legged resonator". The third structure corresponds to the one with the reduced stiffness due to Euler's buckling of the middle rod, which has a relatively large thermal expansion coefficient. This is achieved by applying a temperature of the critical value estimated in (3.6). Dispersion diagrams for the first six eigenfrequencies are produced. Dispersion diagrams have been generated using COMSOL Multiphysics®, together with MATLAB®.

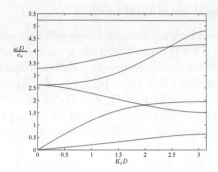

FIGURE 3.3: Dispersion diagram for the doubly periodic frame structure.

For the doubly periodic frame without resonators, a unit square cell is considered with the ratio of the cell wall thickness to the cell width being 1:10. One of the components of the Bloch vector is zero, and the dispersion diagram is produced by varying the other component. The results are shown in Figure 3.3. If the shear wave speed c_s is defined as $c_s = (\mu/\rho)^{1/2}$, then the dimensionless quantity $\omega D/c_s$, is plotted as a function of $K_i D$ where K_i is the non-zero component of the Bloch vector, and D is the unit cell dimension ($D = 1$ in these calculations). In this dimensionless form the only material parameter required is Poisson's ratio ν; other material properties are required for the dimensioned eigenfrequency dispersion diagram. For these calculations, the value $\nu = 0.3$ was used. The essential behaviour is very similar for other values of ν. In view of the Bloch–Floquet condition, the results are symmetric about the line $K_i D = \pi$ and periodic with period 2π. We also note that there are no stop bands adjacent to the origin.

The three-legged resonator is now introduced into the unit cell as shown in Figure 3.1. Each leg has length 0.2 m and width 0.02 m. The resonator block is 0.3 m by 0.14 m and the whole is contained in the unit square frame with $\nu = 0.3$. In this case there are two dispersion diagrams, each formed by one of the components of the Bloch vector being zero and other component varying. The variation of the dimensionless quantity $\omega D/c_s$, is plotted as a function of $K_1 D$ with $K_2 = 0$ and as a function of $K_2 D$ with $K_2 = 0$ in Figure 3.4 (a) and (b), respectively.

Once the temperature of the frame and resonator is above the critical temperature for buckling of the middle rod, the dispersion diagram will change, since the stiffness of the structure will be reduced as discussed earlier. For the two-legged resonator the results are shown as dashed lines in Figure 3.4. The dimensions and conditions are identical to those for the three-legged resonator shown in Figure 3.2. With the introduction of each of the two-legged and three-legged resonators, stop bands appear at higher frequencies. One of the larger stop bands is shown in Figure 3.4b. There is also a low frequency standing wave in Figure 3.4 (a) and (b) just below the value $\omega D/c_s = 0.5$.

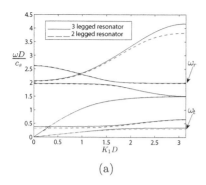
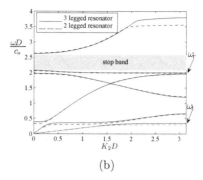

(a)   (b)

FIGURE 3.4: Variation of $\omega D/c_s$ with (a) $K_1 D$ ($K_2 = 0$) and (b) $K_2 D$ ($K_1 = 0$) for the frame with each resonator.

Corresponding results were found for two other values of Poisson's ratio: $\nu = 0.1$ and $\nu = 0.4$. Similar behaviour was found in these results as in those for $\nu = 0.3$.

The effect can be made even more pronounced by constructing the three-legged resonator with a middle leg of larger stiffness than the outer legs, greater differences in the results may be seen. This concept may be extended to a resonator with many more legs between the inner and outer legs. This will increase the effective stiffness of the interior legs resulting in greater differences in the dispersion diagrams before and after buckling. The concept may be further extended by constructing these interior legs of differing thermal expansion coefficients. Selective buckling may be controlled by appropriate choice of the thermal expansion coefficients of the legs with appropriate more gradual change in the dispersion diagrams.

The dispersion surfaces are shown in Figure 3.5 for the frame alone without any resonator and with $\nu = 0.3$.

With the inclusion of the three-legged resonator, the corresponding diagram is shown in Figure 3.6a. In reorientating this figure, as shown in Figure 3.6b, we can see the dispersion surfaces corresponding to the standing waves at frequencies ω_t and ω_r estimated in Equations (3.7) and (3.10), respectively. This diagram also shows a narrow stop band adjacent to the frequency ω_r of the rotational mode.

Constant frequency contours, or "slowness contours", are taken at various dimensionless frequency values for both the frame only and for the frame with three-legged resonator. The contour level for dimensionless frequency value 1 is shown in Figure 3.7a for the frame and Figure 3.7b for the frame and resonator. Corresponding results are shown for contour values 2 and 3 in Figures 3.8 and 3.9 respectively.

In Figure 3.7, we can see a square and rectangular contour corresponding to the acoustic dispersion curves adjacent to the origin. The orientation of the normal to this contour shows sharply defined directions (along the K_1 and K_2

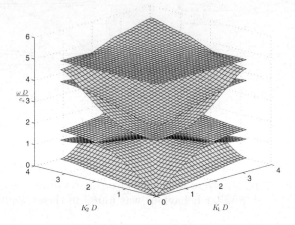

FIGURE 3.5: Dimensionless dispersion surfaces for the frame only; $\nu = 0.3$.

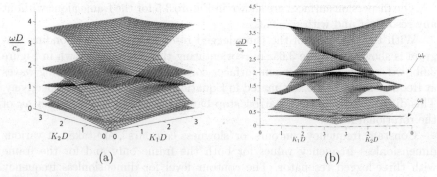

FIGURE 3.6: Dimensionless dispersion surfaces for the frame with three-legged resonator for two different orientations; $\nu = 0.3$.

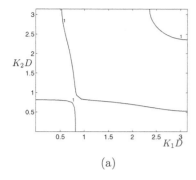 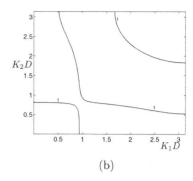

(a)   (b)

FIGURE 3.7: Dimensionless slowness contours, $\omega D/c_s = 1$, for the frame only (a) and the frame with resonator (b); $\nu = 0.3$.

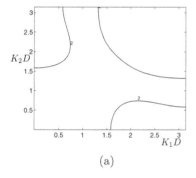 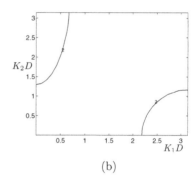

(a)   (b)

FIGURE 3.8: Dimensionless slowness contours, $\omega D/c_s = 2$, for the frame only (a) and the frame with resonator (b); $\nu = 0.3$.

axes) for the effective group velocity and hence the direction of propagation of energy. The presence of the resonator in the figures (b) breaks the symmetry which makes the K_2 direction preferable in terms of transmission of energy for waves of large wavelengths.

The dimensionless frequency values for the translational and rotational eigenmodes, discussed earlier, were calculated using Equations (3.7) and (3.10), respectively. The analytical calculations gave $\omega_t D/c_s = 0.48$ and $\omega_r D/c_s = 2.21$. These agree reasonably with the values from the numerical calculations of $\omega_t D/c_s = 0.38$ and $\omega_r D/c_s = 2.00$ when $K_1 = K_2 = 0$.

We have demonstrated that the position of the stop bands can be controlled via a pre-stress, introduced by applying a temperature. In this case, the boundaries of the stop bands can be determined analytically using asymptotic estimates.

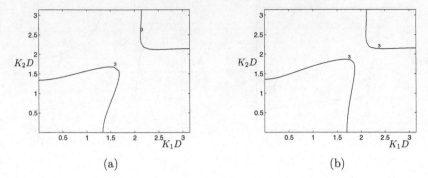

FIGURE 3.9: Dimensionless slowness contours, $\omega D/c_s = 3$, for the frame only (a) and the frame with resonator (b); $\nu = 0.3$.

3.2 Singular perturbation analysis of fields in solids with disintegrating junctions

In practice, many structural elements contain defects which either have been present throughout the component life because of the manufacturing process, or have arisen during the life of the component. Small defects may not affect structural performance greatly and we will concentrate on defects with small ligaments. We will examine models for the junction conditions relevant to multi-scale resonators and, in particular, the effects on their modes of vibration, as in the paper [69].

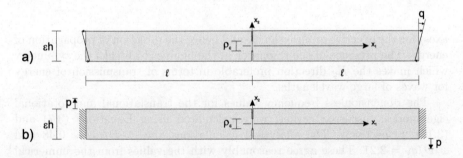

FIGURE 3.10: Elastic domains and boundary conditions (a) for the bending problem and (b) for the shear problem.

An asymptotic model of a junction within a thin-walled elastic solid subjected to a general load will now be addressed. The configuration in question involves a thin body, weakened by two surface breaking cracks, growing toward each other, whose vertices are close to touching, as shown in Figure 3.10. Although it may be tempting to think that small-ligament analysis may not be of great use in practice, such investigation may be important in so-called "leak before break" safety case analyses in engineering where interest is focussed on deep cracks in components.

There are few results when the bridge between the cracks is thin, but for small surface breaking cracks several engineering approaches have been developed to estimate the effective stiffness across the damaged section (see, for example, [34], [185]). In addition, the asymptotic analysis of boundary value problems for domains with thin bridges and non-smooth boundaries is presented in [41] and [109].

A related class of asymptotic problems for thin interfaces includes high-contrast formulations, where the stiffness of the interface layer is defined as a second small parameter, in addition to the small thickness. The asymptotic derivation of a model of adhesively bonded joints was published in [86], while further work involving classification of the lower-dimensional models of adhesively bonded plates was published in [87] and [127]. This includes the derivation of a system of equations, which couples the longitudinal and transverse modes for different cases of the elastic contrast versus the non-dimensional thickness. The derivation and analysis of transmission conditions for thin elastic plates bonded in their common plane by a weak adhesive layer, are given in [70] and [71]. Further extension of the analysis of adhesive joints was developed for the case of orthotropic highly inhomogeneous layered structures in [7], while stability analyses of stratified media with imperfect interfaces have been carried out in [19, 22]. The asymptotic model for a crack, whose faces are joined by elastic fibres, presenting the effective behaviour of the bridged crack is discussed in [1]. Asymptotic and numerical studies for cracks on imperfect interfaces, including structures bonded by weak adhesives, are discussed in [4].

The analysis of tensile junctions in thin-walled truss structures was published in [188, 189]. An important feature of elastic junctions of this type is the discontinuity of the leading-order terms in the asymptotic representation of components of the displacement or their derivatives. One can also refer to such junctions as "spring-like" junctions, where an effective stiffness, either tensile or flexural, can be determined.

The junction conditions are derived via analysis of the boundary layers in the spirit of the asymptotic approach [90]. This analysis required the use of weight functions, which represent special solutions of homogenous model problems in unbounded domains with polynomial growth at infinity.

We will concentrate on effective, spring-like, junction conditions for flexural junctions. The results of the asymptotic analysis will be shown to agree well with finite element numerical computations carried out in the framework of plane strain. For a finite width of the ligament between the vertices of surface-breaking cracks, the numerical results agree well with the estimate presented in [153].

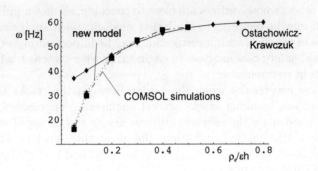

FIGURE 3.11: First radian eigenfrequency ω of a simply supported plate (plane strain) weakened by a pair of surface-breaking cracks in the middle section. The new model refers to the results discussed in this section; the other "Ostachowicz-Krawczuk" model is described in [153].

Here we formulate the bending problem and describe the corresponding polynomial solutions of the Neumann problem for the Lamé system in the infinite strip. The asymptotic procedure requires boundary layer formulations in unbounded domains, followed by the evaluation of the coefficients in the asymptotics at infinity together with the weight functions. Special attention is paid to the derivation of the effective flexural junction condition for the case where two surface-breaking cracks are close to touching. In particular, we discuss the asymptotic approximation of physical fields around a junction subjected to shear and show that the effective shearing stiffness is relatively high unless the thin ligament is exponentially small.

Figure 3.11 shows the computation of the first eigenfrequency of a simply supported plate weakened by a pair of surface-breaking cracks in the middle section: the analytical asymptotic approach agrees well with a finite element computation in COMSOL Multiphysics®, and it also covers the range of geometric parameters previously inaccessible to the existing analytical approximations, discussed in [153]. The problem is extended in section 3.2.2 to the case of a shearing load.

3.2.1 Bending problem

The elastic thin strip domain corresponds to a rectangle of length and height equal to 2ℓ and εh, respectively, with the cross-section in the middle significantly reduced by the presence of two symmetric cracks. The thickness of the bridge between the cracks is denoted by ρ_ε, as illustrated in Figure 3.10.

For the bending problem we apply symmetric rotations to the two ends (at $x_1 = \pm \ell$, Figure 3.10a). Because of symmetry, we may consider only half of the domain, namely $\Omega_\varepsilon = \{\mathbf{x} = (x_1, x_2) \in \mathbb{R}^2 : |0 < x_1 < \ell, |x_2| < \varepsilon h/2\}$ with appropriate boundary conditions. In other words, the solution in terms of displacements is a vector field $\mathbf{u}_\varepsilon(\mathbf{x})$ satisfying the boundary value problem

$$\mu\nabla^2\mathbf{u}_\varepsilon(\mathbf{x}) + (\lambda+\mu)\nabla\nabla\cdot\mathbf{u}_\varepsilon(\mathbf{x}) = 0, \quad \mathbf{x}\in\Omega_\varepsilon, \tag{3.11}$$

$$\sigma_{22}\left(x_1, \pm\frac{\varepsilon h}{2}\right) = \sigma_{12}\left(x_1, \pm\frac{\varepsilon h}{2}\right) = 0, \quad 0 < x_1 < \ell, \tag{3.12}$$

$$u_{\varepsilon 1}(0, x_2) = \sigma_{12}(0, x_2) = 0, \quad |x_2| < \frac{\rho_\varepsilon}{2}, \tag{3.13}$$

$$\sigma_{11}(0, x_2) = \sigma_{12}(0, x_2) = 0, \quad \frac{\rho_\varepsilon}{2} < |x_2| < \frac{\varepsilon h}{2} \tag{3.14}$$

$$\mathbf{u}_\varepsilon(\ell, x_2) = \begin{pmatrix} qx_2 \\ 0 \end{pmatrix}, \quad -\frac{\varepsilon h}{2} < x_2 < \frac{\varepsilon h}{2}, \tag{3.15}$$

where q is the applied rotation.

If we introduce the scaled variable $\xi_2 = x_2/\varepsilon$, the boundary value problem (3.11)–(3.15) scales accordingly with the length of the thin bridge in the scaled coordinates given by $c_\varepsilon = \rho_\varepsilon/\varepsilon$. For the bending problem, the vector $\mathbf{u}_\varepsilon(\mathbf{x})$ admits the representation [109, 127]

$$\mathbf{u}_\varepsilon \sim \varepsilon^{-2}\mathbf{u}^{(0)} + \varepsilon^{-1}\mathbf{u}^{(1)} + \mathbf{u}^{(2)} + \varepsilon\mathbf{u}^{(3)} + \varepsilon^2\mathbf{u} + \varepsilon^{-2}\mathbf{L}^{0,\ell}, \tag{3.16}$$

where the vector functions $\mathbf{u}^{(i)}$ ($i = 0,\ldots,3$) and $\mathbf{u}$ depend on (x_1, ξ_2) and the boundary-layer terms, $\mathbf{L}^0$ (at $x_1 = 0$) and $\mathbf{L}^\ell$ (at $x_1 = \ell$), decay exponentially far from the side where they are constructed. We can show (see [127]) that the leading order term $\mathbf{u}^{(0)}$ is of the form $(0, u_2^{(0)}(x_1))^T$, where $u_2^{(0)}(x_1)$ obeys the ordinary differential equation

$$\frac{h^3}{12}\frac{4\mu(\lambda+\mu)}{(2\mu+\lambda)}u_2^{(0)''''}(x_1) = 0, \tag{3.17}$$

with $4\mu(\lambda+\mu)/(2\mu+\lambda) = E/(1-\nu^2)$, and the remaining terms $\mathbf{u}^{(j)}$ ($j = 1,2,3$) are

$$\mathbf{u}^{(1)}(x_1,\xi_2) = \begin{pmatrix} -\xi_2 u_2^{(0)'}(x_1) \\ 0 \end{pmatrix}, \tag{3.18}$$

$$\mathbf{u}^{(2)}(x_1,\xi_2) = \begin{pmatrix} 0 \\ \dfrac{\lambda}{2(2\mu+\lambda)}\left(\xi_2^2 - \dfrac{h^2}{12}\right)u_2^{(0)''}(x_1) \end{pmatrix}, \tag{3.19}$$

$$\mathbf{u}^{(3)}(x_1,\xi_2) = \begin{pmatrix} \dfrac{1}{6}\left[\dfrac{3\lambda+4\mu}{\lambda+2\mu}\xi_2^3 - \dfrac{11\lambda+12\mu}{4(\lambda+2\mu)}\xi_2 h^2\right]u_2^{(0)'''}(x_1) \\ 0 \end{pmatrix}. \tag{3.20}$$

This particular form of $\mathbf{u}^{(2)}$ and $\mathbf{u}^{(3)}$ has been chosen in order to have zero average across the thickness.

It is noted that in the article [20], the representation (3.16) has been generalised to the case of a longitudinal pre-stress in the plate.

Equation (3.17) requires four boundary conditions that can be obtained by investigating the boundary layers at $x_1 = 0$ and $x_1 = \ell$. On the right-hand side of the strip ($x_1 = \ell$) the procedure is well-known, providing the classical results consistent with the Saint–Venant principle

$$u_2^{(0)}(\ell) = 0, \quad u_2^{(0)\prime}(\ell) = -q. \tag{3.21}$$

An exponential boundary layer occurs at the left end of the strip.

3.2.1.1 Boundary layer at the junction

To define the boundary layer in the vicinity of the damaged section, let us introduce the scaled coordinate $\xi_1 = x_1/\varepsilon$. The boundary layer $\mathbf{L}^0(\boldsymbol{\xi})$ satisfies the homogeneous Lamé system in the infinite strip $\Pi_+ = \{\boldsymbol{\xi} = (\xi_1, \xi_2) : \xi_1 > 0, |\xi_2| < h/2\}$. For the sake of convenience, we shall omit the superscript ^{0} in the formulae below. Employing (3.16), to the first order, the displacements and stress components at the left-hand boundary of Π_+ ($\xi_1 = 0$) turn out to be

$$u_1(0, \xi_2) = -\frac{1}{\varepsilon} \xi_2 \, u_2^{(0)\prime}(0) + \frac{1}{\varepsilon^2} L_1(0, \xi_2), \tag{3.22}$$

$$u_2(0, \xi_2) = \frac{1}{\varepsilon^2} u_2^{(0)}(0) + \frac{1}{\varepsilon^2} L_2(0, \xi_2), \tag{3.23}$$

$$\sigma_{11}^u(0, \xi_2) = \frac{1}{\varepsilon} \frac{4\mu(\lambda + \mu)}{\lambda + 2\mu} \xi_2 u_2^{(0)\prime\prime}(0) + \frac{1}{\varepsilon^3} \sigma_{11}^L(0, \xi_2), \tag{3.24}$$

$$\sigma_{12}^u(0, \xi_2) = \frac{\mu(\lambda + \mu)}{2(\lambda + 2\mu)} (4\xi_2^2 - h^2) u_2^{(0)\prime\prime\prime}(0) + \frac{1}{\varepsilon^3} \sigma_{12}^L(0, \xi_2), \tag{3.25}$$

where $\sigma^u(x_1, \xi_2) = \lambda(\mathrm{tr}\, \nabla \mathbf{u}_\varepsilon) \mathbf{I} + \mu(\nabla \mathbf{u}_\varepsilon + \nabla \mathbf{u}_\varepsilon^T)$ with $\mathbf{u}_\varepsilon$ given by (3.16), and

$$\sigma_{11}^L(\boldsymbol{\xi}) = (\lambda + 2\mu) \frac{\partial L_1}{\partial \xi_1} + \lambda \frac{\partial L_2}{\partial \xi_2}, \quad \sigma_{12}^L(\boldsymbol{\xi}) = \mu\left(\frac{\partial L_1}{\partial \xi_2} + \frac{\partial L_2}{\partial \xi_1}\right). \tag{3.26}$$

Therefore, to match Equations (3.13) and (3.14), the boundary conditions for the boundary-layer term $\mathbf{L}$ are

$$\left.\begin{aligned} L_1(0, \xi_2) &= \varepsilon \xi_2 u_2^{(0)\prime}(0) \\ \sigma_{12}^L(0, \xi_2) &= -\varepsilon^3 \gamma \frac{4\xi_2^2 - h^2}{4} u_2^{(0)\prime\prime\prime}(0) \end{aligned}\right\} \quad |\xi_2| < \frac{c_\varepsilon}{2}, \tag{3.27}$$

$$\left.\begin{aligned} \sigma_{11}^L(0, \xi_2) &= \varepsilon^2 2\gamma \xi_2 u_2^{(0)\prime\prime}(0) \\ \sigma_{12}^L(0, \xi_2) &= -\varepsilon^3 \gamma \frac{4\xi_2^2 - h^2}{4} u_2^{(0)\prime\prime\prime}(0) \end{aligned}\right\} \quad \frac{c_\varepsilon}{2} < |\xi_2| < \frac{h}{2}, \tag{3.28}$$

$$\sigma_{12}^L\left(\xi_1, \pm\frac{h}{2}\right) = \sigma_{22}^L\left(\xi_1, \pm\frac{h}{2}\right) = 0, \quad \xi_1 > 0, \tag{3.29}$$

where
$$\gamma = \frac{2\mu(\lambda+\mu)}{\lambda+2\mu}. \tag{3.30}$$

The boundary layer does not introduce an error into the boundary conditions (3.13) and (3.14) if
$$u_2^{(0)\prime\prime\prime}(0) = 0, \tag{3.31}$$

so that the right-hand sides in the second condition of (3.27) and the second condition in (3.28) become zero. In this case Equations (3.27) and (3.28) reduce to the form

$$\left. \begin{array}{l} L_1(0,\xi_2) = \varepsilon\xi_2 u_2^{(0)\prime}(0) \\ \sigma_{12}^L(0,\xi_2) = 0 \end{array} \right\} \quad |\xi_2| < \frac{c_\varepsilon}{2}, \tag{3.32}$$

$$\left. \begin{array}{l} \sigma_{11}^L(0,\xi_2) = \varepsilon^2 2\gamma\xi_2 u_2^{(0)\prime\prime\prime}(0) \\ \sigma_{12}^L(0,\xi_2) = 0 \end{array} \right\} \quad \frac{c_\varepsilon}{2} < |\xi_2| < \frac{h}{2}. \tag{3.33}$$

3.2.1.2 Weight function and the junction condition

To derive the effective junction condition at $x_1 = 0$ we require the weight function, defined as a special solution of the homogenous boundary-value problem in Π_+, with the polynomial growth at infinity. For the bending problem, the relevant weight function $\mathbf{Z}(\boldsymbol{\xi})$ is defined as the solution of the following problem:

$$\mu\nabla^2\mathbf{Z}(\boldsymbol{\xi}) + (\lambda+\mu)\nabla\nabla\cdot\mathbf{Z}(\boldsymbol{\xi}) = \mathbf{0}, \quad \boldsymbol{\xi} \in \Pi_+, \tag{3.34}$$

$$\sigma_{22}^Z\left(\xi_1, \pm\frac{h}{2}\right) = \sigma_{12}^Z\left(\xi_1, \pm\frac{h}{2}\right) = 0, \quad \xi_1 > 0, \tag{3.35}$$

$$Z_1(0,\xi_2) = \sigma_{12}^Z(0,\xi_2) = 0, \quad |\xi_2| < \frac{c_\varepsilon}{2}, \tag{3.36}$$

$$\sigma_{11}^Z(0,\xi_2) = \sigma_{12}^Z(0,\xi_2) = 0, \quad \frac{c_\varepsilon}{2} < |\xi_2| < \frac{h}{2}, \tag{3.37}$$

where $\sigma_{ij}^Z(\boldsymbol{\xi})$ ($i,j = 1,2$) are defined as in (3.26) with functions $Z_1(\boldsymbol{\xi})$, $Z_2(\boldsymbol{\xi})$ replacing $L_1(\boldsymbol{\xi})$, $L_2(\boldsymbol{\xi})$. Moreover,

$$\mathbf{Z}(\boldsymbol{\xi}) = \begin{pmatrix} \dfrac{\xi_1\xi_2}{h} \\ -\dfrac{1}{2h}\left(\xi_1^2 + \dfrac{\lambda}{\lambda+2\mu}\xi_2^2\right) \end{pmatrix} + \hat{\mathbf{Z}}(\boldsymbol{\xi}), \quad \text{as } \xi_1 \to +\infty, \tag{3.38}$$

where $\hat{\mathbf{Z}}$ is bounded. This gives a stress distribution

$$\sigma_{11}^Z = 2\gamma\frac{\xi_2}{h}, \quad \sigma_{22}^Z = \sigma_{12}^Z = 0, \quad \text{as } \xi_1 \to +\infty, \tag{3.39}$$

that is equivalent to a bending moment $M = \gamma h^2/6$ (clockwise).

Application of the Betti formula to **L** and **Z** (integration by parts) yields

$$u_2^{(0)\prime}(0) \int_{-c_\varepsilon/2}^{c_\varepsilon/2} \xi_2 \sigma_{11}^Z(0,\xi_2)\, d\xi_2$$

$$-2\varepsilon\gamma u_2^{(0)\prime\prime\prime}(0) \left[\int_{-h/2}^{-c_\varepsilon/2} \xi_2 Z_1(0,\xi_2)\, d\xi_2 + \int_{c_\varepsilon/2}^{h/2} \xi_2 Z_1(0,\xi_2)\, d\xi_2 \right] = 0. \quad (3.40)$$

Equilibrium requires that the resultant moment of stress $\sigma_{11}^Z(0,\xi_2)$ in the thin bridge ($\xi_1 = 0, |\xi_2| < c_\varepsilon/2$), namely the first integral in (3.40), be equal to $\gamma h^2/6$ (counter-clockwise), then condition (3.40) becomes

$$u_2^{(0)\prime}(0) h^2 - 24\, \varepsilon u_2^{(0)\prime\prime\prime}(0) \int_{c_\varepsilon/2}^{h/2} \xi_2 Z_1(0,\xi_2)\, d\xi_2 = 0. \quad (3.41)$$

To evaluate the integral in (3.41), we obtain an estimate of $Z_1(0,\xi_2)$ for $\xi_2 > 0$. Anti-symmetry with respect to the longitudinal axis of the applied load as $\xi_1 \to \infty$ (see (3.38) and (3.39)) and boundary conditions (3.36) and (3.37) suggest that the weight function has the asymptotic form corresponding to the rigid rotation of the vertical boundary. Hence $Z_1(0,\xi_2)$ can be approximated as $Z_1(0,\xi_2) \sim \eta(\xi_2 - c_\varepsilon/2)$, where η is a rotation that can be evaluated as follows. Consider the situation where the domain Π_+ is in equilibrium by means of the two moments M described above, one as $\xi_1 \to \infty$ and the other at the origin (Figure 3.12). At the latter point, we can separate half a circle of radius equal to $c_\varepsilon/2$ from the semi-infinite strip. The rotation needed can be deduced as the difference between the rotation of the edge AB evaluated at points B and A, respectively.

This difference can be obtained employing the analysis shown in [32], where the rotation of an elastic wedge (with a generic opening angle) subjected to a concentrated moment at the tip is reported. For this problem, at a distance r from the tip, the rotational stiffness is given by $4\mu D r^2$, where D is a coefficient depending on the opening angle and on the Poisson's ratio. In plane strain, an approximation of D when the wedge becomes a half space, is

$$D = \frac{\pi}{5 - 2\nu}. \quad (3.42)$$

Thus, the approximate expression of the function $Z_1(0,\xi_2)$ is given by

$$Z_1(0,\xi_2) \sim \frac{M}{4\mu D(c_\varepsilon/2)^2} \left(\xi_2 - \frac{c_\varepsilon}{2} \right), \quad (3.43)$$

so that the remaining integral in (3.41) can be evaluated to give, to the first order, the *effective junction condition*

$$u_2^{(0)\prime}(0) - \frac{5 - 2\nu}{6\pi(1 - \nu)} \frac{(\varepsilon h)^3}{\rho_\varepsilon^2} u_2^{(0)\prime\prime\prime}(0) = 0. \quad (3.44)$$

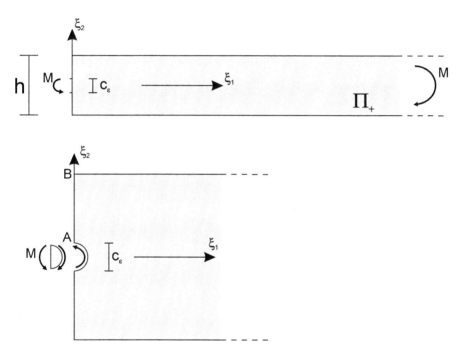

FIGURE 3.12: Estimate of the weight function $Z_1(0, \xi_2)$.

Condition (3.44) can be reformulated in terms of *effective elastic spring stiffness* K_b employing the constitutive law

$$\mathcal{M} = u_2^{(0)''}(0) \frac{E(\varepsilon h)^3}{12(1-\nu^2)}$$

between the bending moment $\mathcal{M}$ and curvature of the strip $u_2^{(0)''}(0)$. The elastic response at the junction, where a jump in rotation $\Delta\phi$ equal to $2u_2^{(0)'}(0)$ is found, can be represented as

$$\mathcal{M} = K_b \Delta\phi, \qquad (3.45)$$

with

$$K_b = \frac{\pi E \rho_\varepsilon^2}{4(5-2\nu)(1+\nu)}. \qquad (3.46)$$

The coefficient in (3.44) of $u_2^{(0)'''}(0)$ is of order $O(1)$ provided $\rho_\varepsilon = A\varepsilon^{\frac{3}{2}}$, where A is a constant.

3.2.2 Shear problem

For the same domain Ω_ε, we consider the following problem (see Figure 3.10b):

$$\mu \nabla^2 \mathbf{u}_\varepsilon(\mathbf{x}) + (\lambda + \mu)\nabla\nabla \cdot \mathbf{u}_\varepsilon(\mathbf{x}) = \mathbf{0}, \quad \mathbf{x} \in \Omega_\varepsilon, \qquad (3.47)$$

$$\sigma_{22}\left(x_1, \pm\frac{\varepsilon h}{2}\right) = \sigma_{12}\left(x_1, \pm\frac{\varepsilon h}{2}\right) = 0, \quad 0 < x_1 < \ell, \qquad (3.48)$$

$$\sigma_{11}(0, x_2) = u_{\varepsilon 2}(0, x_2) = 0, \quad |x_2| < \frac{\rho_\varepsilon}{2}, \qquad (3.49)$$

$$\sigma_{11}(0, x_2) = \sigma_{12}(0, x_2) = 0, \quad \frac{\rho_\varepsilon}{2} < |x_2| < \frac{\varepsilon h}{2}, \qquad (3.50)$$

$$\mathbf{u}_\varepsilon(\ell, x_2) = \begin{pmatrix} 0 \\ -p \end{pmatrix}, \quad -\frac{\varepsilon h}{2} < x_2 < \frac{\varepsilon h}{2}, \qquad (3.51)$$

where p is constant.

The asymptotic representation (3.16) is replaced by

$$\mathbf{u}_\varepsilon \sim \varepsilon^{-2}\mathbf{u}^{(0)} + \varepsilon^{-1}\mathbf{u}^{(1)} + \mathbf{u}^{(2)} + \varepsilon\mathbf{u}^{(3)} + \varepsilon^2\mathbf{u} + \varepsilon^{-2}\mathbf{H}, \qquad (3.52)$$

where $\mathbf{H}(\boldsymbol{\xi})$ is the new boundary layer for the junction subjected to the shear load. Following the Taylor expansion of $u_{\varepsilon 1}$, $u_{\varepsilon 2}$ given by (3.18)–(3.20) we obtain expressions (3.22)–(3.25) with $\mathbf{L}^0$ being replaced by $\mathbf{H}$. Near the junction, where $\mathbf{H}$ satisfies the homogeneous Lamé system, the boundary conditions are

$$\left. \begin{array}{l} \sigma_{11}^H(0, \xi_2) = \varepsilon^2 2\gamma\, \xi_2 u_2^{(0)'''}(0) \\[4pt] H_2(0, \xi_2) = -u_2^{(0)}(0) \end{array} \right\} \quad 0 < |\xi_2| < \frac{c_\varepsilon}{2}, \qquad (3.53)$$

$$\left.\begin{array}{l}\sigma_{11}^H(0,\xi_2) = \varepsilon^2 2\gamma\,\xi_2 u_2^{(0)\prime\prime}(0)\\[4pt]\sigma_{12}^H(0,\xi_2) = -\varepsilon^3\gamma\dfrac{4\xi_2^2 - h^2}{4}u_2^{(0)\prime\prime\prime}(0)\end{array}\right\}\quad \dfrac{c_\varepsilon}{2} < |\xi_2| < \dfrac{h}{2}, \qquad (3.54)$$

$$\sigma_{12}^H\left(\xi_1,\pm\dfrac{h}{2}\right) = \sigma_{22}^H\left(\xi_1,\pm\dfrac{h}{2}\right) = 0,\quad \xi_1 > 0. \qquad (3.55)$$

We would like to construct the boundary layer $\mathbf{H}$ in such a way that it does not introduce an error in the boundary conditions (3.49) and (3.50) for σ_{11}, and in this case $u_2^{(0)\prime\prime}(0) = 0$. Then it is possible, by means of a weight function, to obtain a junction condition relating $u_2^{(0)}$ to $u_2^{(0)\prime\prime\prime}$, or equivalently, the displacement jump across the junction to the shear force. On the right-hand side of the strip ($x_1 = \ell$) the standard procedure provides the additional two boundary conditions for Equation (3.17), namely

$$u_2^{(0)}(\ell) = -p,\quad u_2^{(0)\prime}(\ell) = 0. \qquad (3.56)$$

3.2.2.1 Representation of the junction condition in terms of the weight function

A new weight function, say $\mathbf{W}(\boldsymbol{\xi})$, is required for the derivation of the effective junction condition for the case of the external shear load. This is defined as a solution of the problem

$$\mu\nabla^2\mathbf{W}(\boldsymbol{\xi}) + (\lambda+\mu)\nabla\nabla\cdot\mathbf{W}(\boldsymbol{\xi}) = \mathbf{0},\quad \boldsymbol{\xi}\in\Pi_+, \qquad (3.57)$$

$$\sigma_{22}^W\left(\xi_1,\pm\dfrac{h}{2}\right) = \sigma_{12}^W\left(\xi_1,\pm\dfrac{h}{2}\right) = 0,\quad \xi_1 > 0, \qquad (3.58)$$

$$W_2(0,\xi_2) = \sigma_{11}^W(0,\xi_2) = 0,\quad |\xi_2| < \dfrac{c_\varepsilon}{2}, \qquad (3.59)$$

$$\sigma_{11}^W(0,\xi_2) = \sigma_{12}^W(0,\xi_2) = 0,\quad \dfrac{c_\varepsilon}{2} < |\xi_2| < \dfrac{h}{2}, \qquad (3.60)$$

with

$$\mathbf{W}(\boldsymbol{\xi}) = \begin{pmatrix} \dfrac{4\mu+3\lambda}{2\mu+\lambda}\dfrac{\xi_2^3}{3h^2} - \dfrac{\xi_1^2\xi_2}{h^2} - \dfrac{\mu+\lambda}{2\mu+\lambda}\xi_2 \\[8pt] \dfrac{\xi_1}{h^2}\left(\dfrac{\xi_1^2}{3} + \dfrac{\lambda}{\lambda+2\mu}\xi_2^2\right) \end{pmatrix} + \hat{\mathbf{W}}(\boldsymbol{\xi}),\quad \text{as } \xi_1\to +\infty, \qquad (3.61)$$

where $\hat{\mathbf{W}}$ is bounded. This gives the stress distribution

$$\sigma_{11}^W = -4\gamma\dfrac{\xi_1\xi_2}{h^2},\quad \sigma_{12}^W = \dfrac{\gamma}{2h^2}(4\xi_2^2 - h^2),\quad \sigma_{22}^W = 0,\quad \text{as } \xi_1\to +\infty, \qquad (3.62)$$

which produces a constant shear force $T = -\gamma h/3$.

Application of the Betti formula to the fields $\mathbf{H}$ and $\mathbf{W}$ yields

$$-u_2^{(0)}(0) \int_{-c_\varepsilon/2}^{c_\varepsilon/2} \sigma_{12}^W(0, \xi_2) \, d\xi_2$$

$$+ \varepsilon^3 \frac{\gamma u_2^{(0)\prime\prime\prime}(0)}{4} \left[\int_{-h/2}^{-c_\varepsilon/2} (4\xi_2^2 - h^2) W_2(0, \xi_2) \, d\xi_2 \right.$$

$$\left. + \int_{c_\varepsilon/2}^{h/2} (4\xi_2^2 - h^2) W_2(0, \xi_2) \, d\xi_2 \right] = 0. \quad (3.63)$$

Equilibrium requires that the resultant shear force in the thin bridge ($\xi_1 = 0, |\xi_2| < c_\varepsilon/2$), namely the first integral in (3.63), be equal to $\gamma h/3$, then Equation (3.63) becomes

$$-u_2^{(0)}(0)h + \frac{3}{4} \varepsilon^3 u_2^{(0)\prime\prime\prime}(0) \beta(c_\varepsilon) = 0, \quad (3.64)$$

where $\beta(c_\varepsilon)$ represents the integrals in brackets in Equation (3.63).

3.2.2.2 Effective stiffness of the junction

The function $W_2(0, \xi_2)$ is written as

$$W_2(0, \xi_2) = -\frac{h}{3\pi} \log \left| \frac{2\xi_2}{c_\varepsilon} \right| + V(\xi_2), \quad (3.65)$$

where $V(\xi_2)$ is a bounded function, and the first term in Equation (3.65) corresponds to the vertical component of the displacement produced by the tangential force applied to the boundary of the half plane $\xi_1 > 0$ vanishing at $\xi_2 = c_\varepsilon/2$. Similar double boundary layer structures occur in problems related to tensile loading of crack junctions in [188, 189].

The substitution of (3.65) in (3.64) allows us to calculate the integrals, from which it turns out that, to the leading order, the junction condition in the shear problem is

$$-u_2^{(0)}(0) + \frac{(\varepsilon h)^3}{6\pi} u_2^{(0)\prime\prime\prime}(0) \log \frac{\varepsilon h}{\rho_\varepsilon} = 0. \quad (3.66)$$

This can be reformulated in terms of relationship between the shear force applied at the junction ($\mathcal{T}$) and the displacement jump $\partial u = 2u_2^{(0)}(0)$ as $\mathcal{T} = K_s \partial u$. To this end, it is necessary to employ the constitutive law

$$\mathcal{T} = u_2^{(0)\prime\prime\prime}(0) \frac{E(\varepsilon h)^3}{12(1 - \nu^2)},$$

which yields

$$K_s = \frac{\pi E}{4(1 - \nu^2)} \frac{1}{\log \frac{\varepsilon h}{\rho_\varepsilon}}. \quad (3.67)$$

The coefficient in Equation (3.66) of $u_2^{(0)\prime\prime\prime}(0)$ is of the order $O(1)$ provided $\rho_\varepsilon = B\exp(-\varepsilon^{-3})$, where B is a constant, which implies that we have the spring-like junction if the thickness of the bridge is exponentially small. Alternatively, the displacements are continuous across the junction to the leading-order approximation.

3.2.3 Comparison with other models

Finite element computations have been performed for a thin disintegrating solid shown in Figure 3.10 under plane strain conditions. The solid is weakened by a pair of surface-breaking cracks in the middle section. The COMSOL Multiphysics® and the MATLAB® environments were used for these simulations. The parameter values are: $\ell = 2.5$ m, $\varepsilon h = 0.1$ m, $E = 200$ GPa, and $\nu = 0.3$.

The first angular eigenfrequency ω of the simply supported plate is presented in Figure 3.11 as a function of $\rho_\varepsilon/\varepsilon h$. This graph clearly shows the effectiveness of this analytical approach, which on one hand agrees well with the finite element computation and on the other hand covers the range of parameters inaccessible in the previous analytical models existing in the literature. For example, results of a model by Ostachowicz and Krawczuk [153] are shown which do not agree well for very small ligament lengths.

The effective junction stiffness, given in (3.46), has also been compared with the COMSOL Multiphysics® finite element results. Shear deformation of the junction has been neglected. These results are shown in Figure 3.13, again for varying values of $\rho_\varepsilon/\varepsilon h$. The parameter values and conditions were the same as above.

The law proposed in [153], is

$$K_b^{OK} = \frac{E(\varepsilon h)^2}{36\pi(1-\nu^2)(1-\frac{\rho_\varepsilon}{\varepsilon h})^2 f_D(1-\frac{\rho_\varepsilon}{\varepsilon h})}, \quad (3.68)$$

where the function $f_D(\alpha)$ is

$$f_D(\alpha) = 0.5335 - 0.929\alpha + 3.500\alpha^2 - 3.181\alpha^3 + 5.793\alpha^4. \quad (3.69)$$

It is shown in Figure 3.13a that the proposed model (3.46) agrees well with the simulations for $\rho_\varepsilon/\varepsilon h < 0.35$. For $\rho_\varepsilon/\varepsilon h > 0.35$ the actual response becomes stiffer than our approximate model and shows a good agreement with (3.68), as illustrated in Figure 3.13b.

Two-dimensional spring-like junctions in thin elastic body bodies have been identified in connection with closely positioned surface-breaking cracks when the solids are subjected to bending and shear loads. Related results for junctions, subjected to tensile loads, in truss structures are discussed in [188, 189].

High-order asymptotics of the elastic fields have been constructed within the singularly perturbed region. It is essential to note that these asymptotic approximations include the boundary layers, and the approximations are uniform throughout the entire singularly perturbed domain.

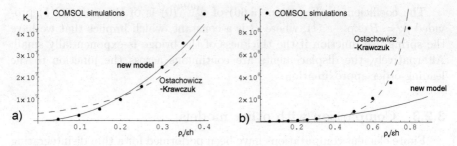

FIGURE 3.13: Comparison between the asymptotic model of bending stiffness (3.46) (solid line), the model of Ostachowicz-Krawczuk [153] (dashed line), and COMSOL Multiphysics® simulations. (a) thin junction range ($\rho_\varepsilon/\varepsilon h < 0.35$); (b) whole range of parameter $\rho_\varepsilon/\varepsilon h$.

3.3 Structures with damaged multi-scale resonators

Here, the asymptotic approach described in Section 3.2 will be applied to the situation when resonators, analysed in Section 3.1, are now damaged.

Waves in discrete lattice structures give a simple and efficient framework for the analysis of dynamic response of micro-structured solids. Fundamental work on dynamic discrete structures was published by Slepyan [178], whose work also included the models of fracture and dissipation within structured media with discrete systems of bonds.

Discrete models have limitations but they are efficient for low frequency regimes. Asymptotic models, combining discrete approximations with the analysis of continuous media, were studied in [189]. Boundary layer solutions were used to derive effective boundary conditions (junction conditions) for thin-walled solids with transverse cracks. Furthermore, these model solutions were used in the analysis of Bloch–Floquet waves in periodic media, in addition to standing waves and localisation in structured solids with defects. For problems of vector elasticity, effective junction conditions for thin solids with cracks were analysed in Section 3.2, where both bending and shear loadings were considered for a junction region containing a transverse crack.

For flexural waves interacting with perforated plate structures, analytical and numerical modelling of apparent negative refraction was presented in [66]. This also includes focussing of flexural waves by a structured interface. For a scalar formulation involving the Helmholtz equation, a thorough study of negative refraction and the corresponding analysis of dispersion properties of Bloch–Floquet waves in doubly periodic structures was published in [110]. For scalar problems of acoustics, which reduce to analysis of solutions of the Helmholtz equation, defect modes and focussing effects are discussed in [76].

In this section, we extend the work presented in Section 3.2 to the case when a continuum periodic structure contains multi-scale resonators of the same nature as in Section 3.1. Analytical asymptotic analysis is applied to individual resonators, whereas the numerical study is performed for Bloch–Floquet waves in an infinite periodic structure.

Asymptotic models of multi-scale resonators lead to explicit analytical approximations of the first few eigenfrequencies and the corresponding eigenmodes. Furthermore, by introducing damage into the resonators, we estimate the change in the eigenfrequencies of the structure.

The analytical work is complemented by numerical simulations, which illustrate specially designed properties of micro-structured solids, such as negative refraction, focussing by a flat structured interface and suppression of the lateral vibrations of a long micro-structured solid.

Section 3.3.1 provides an asymptotic analysis of thin-legged resonators with cracks. Explicit asymptotic approximations of low-frequency vibration modes are derived for out-of-plane shear with cracked regions being replaced by "spring-like" junction conditions. Furthermore, analyses of dispersion diagrams for Bloch–Floquet waves, localisation and focussing in multi-scale periodic structures are presented in detail.

Filtering of out-of-plane shear waves by a structured interface is analysed in Section 3.3.2. A certain range of frequencies is determined for which the structure possesses the property of negative refraction. Both undamaged and damaged resonators are considered. It is shown that significant damage of resonator legs leads to a marginal reduction in frequencies corresponding to negative refraction.

In Section 3.3.3, the models of damage for the multi-scale resonators have been extended to the vector case of plane strain elasticity. Compared with the scalar case, we have flexural and tensile modes of vibration around the junction region with a transverse crack. The analytical approximations of eigenfrequencies of standing waves within micro-structured media are accompanied by the numerical study of dispersion properties of Bloch–Floquet waves. The asymptotic prediction for the fundamental modes of the multi-scale resonators agree well with the results of independent finite element simulations.

In Section 3.3.4, the analytical and numerical results are consolidated into a class of important engineering applications. This involves the demonstration of unusual transmission properties of a micro-structured elastic slab with embedded resonators. Within a selected frequency range, the slab may induce the focussing of an incident wave generated by a point source.

3.3.1 Out-of-plane vibration of a periodic structure with multi-scale resonators

Consider a periodic structure with the elementary cell shown in Figure 3.14a. The geometry of the multi-scale solid with embedded resonators and the corresponding notations were introduced in Section 3.1.1. The out-of-

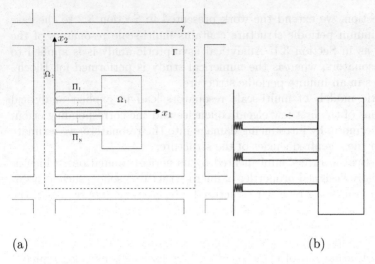

(a)   (b)

FIGURE 3.14: (a) Elementary cell containing multi-scale resonator within a doubly periodic structure. (b) Model of the resonator with 'spring-like' damaged junctions employed in the analysis of the out-of-plane vibration problem.

plane shear Bloch–Floquet waves in such a structure are discussed below. The dispersion diagrams are analysed to show the presence of low frequency standing waves, and asymptotic estimates are derived for their frequencies.

3.3.1.1 Asymptotic approximations for the lowest eigenfrequency

Here we give an asymptotic estimate of the lowest frequency standing wave and relate this to the compliance of the thin rods. Let u denote the amplitude of time-harmonic vibrations of radian frequency ω. Assume that such a vibration corresponds to a standing wave.

For such a mode, the resonator within the elementary cell will be moving with an amplitude much larger than that of the surrounding frame and hence, for a simple estimate, we can assume that amplitude of the frame is negligibly small.

As shown in [90], the leading term of the displacement amplitude in Ω_1 is constant, say A. Assuming that the inertia of the thin rods can be neglected, the displacement amplitudes $u_j(x_1)$ within Π_j can be approximated by linear functions of the local longitudinal variable x_1

$$u_j(x_1) = B_1^{(j)} x_1 + B_2^{(j)}. \tag{3.70}$$

Ideal contact between a thin rod Π_j and the body Ω implies the continuity of the displacement, i.e.

$$u_j(l) = A, \tag{3.71}$$

whereas at the damaged foundation of the thin rod, we have the spring-like

Robin boundary condition:
$$u'_j(0) = \alpha_j u_j(0), \tag{3.72}$$

where α_j represents a measure of the effective stiffness of the damaged junction region. Hence the function u_j takes the form

$$u_j(x_1) = \frac{(1+\alpha_j x_1)A}{(1+\alpha_j l)}. \tag{3.73}$$

Furthermore, the equation of motion for the large solid Ω_1 of mass M, subjected to the forces F_j exerted by thin rods Π_j ($j=1,\ldots,N$) becomes

$$\sum_{j=1}^{N} F_j = -MA\omega^2, \tag{3.74}$$

where
$$F_j = -c_j u'_j(l), \tag{3.75}$$

with c_j being the stiffness of the undamaged rod at the junction region connecting Π_j and Ω_1.

For the sake of simplicity, we assume that all undamaged rods are identical, i.e. $c_j = c$ ($j=1,\ldots,N$). Then it follows from (3.73), (3.74), and (3.75) that the lowest eigenfrequency ω_1^{ap} (the superscript 'ap' indicates 'out-of-plane') for a standing wave is approximated by

$$(\omega_1^{\text{ap}})^2 \simeq \frac{c}{Ml} \sum_{k=1}^{N} \frac{1}{1+\beta_k}, \tag{3.76}$$

where $\beta_k = (\alpha_k l)^{-1}$ is the effective dimensionless compliance of the kth damaged junction.

In addition, we introduce the constraint on the overall effective stiffness of spring-like junctions at the foundations of the thin rods

$$\sum_{k=1}^{N} \alpha_k = \frac{p}{l} = \text{constant}, \tag{3.77}$$

which implies
$$\sum_{k=1}^{N} \frac{1}{\beta_k} = p = \text{constant}. \tag{3.78}$$

We will now show that the asymptotic approximation (3.76) with the constraint (3.77) leads to the conclusion that any non-uniformity in the distribution of damage between the rods will lower the first eigenfrequency of the structure shown in Figure 3.14b, and correspondingly will reduce the frequency of the first standing wave mode for the periodic system in Figure 3.14a.

This can be verified in a standard way. Namely, let

$$f(\beta_1, \ldots, \beta_N) = \sum_{k=1}^{N} \frac{1}{1+\beta_k},$$

subject to the constraint

$$g(\beta_1, \ldots, \beta_N) = 0, \text{ where } g(\beta_1, \ldots, \beta_N) = -p + \sum_{k=1}^{N} \frac{1}{\beta_k}.$$

By considering

$$\phi(\beta_1, \ldots, \beta_N) = f - \lambda g,$$

where λ is the Lagrange multiplier, we identify the extremal configuration by

$$\frac{\partial \phi}{\partial \beta_j} = 0 \ (j = 1, \ldots, N), \quad -p + \sum_{k=1}^{N} \frac{1}{\beta_k} = 0.$$

Eliminating λ, we deduce that

$$\frac{\beta_i^2}{(1+\beta_i)^2} = \frac{\beta_j^2}{(1+\beta_j)^2} \quad (i \neq j, \text{no summation over repeated indices}),$$

which yields, for the case of positive β_j, that

$$\beta_1 = \beta_2 = \ldots = \beta_N = N/p. \tag{3.79}$$

In turn, the maximum value of the function f is

$$f(N/p, \ldots, N/p) = Np/(p+N).$$

A particularly simple illustration is related to the case when $N = 2$. In this situation,

$$(\omega_1^{\text{ap}})^2 \simeq \frac{c}{Ml}\left(\frac{1}{1+\beta_1} + \frac{1}{1+\beta_1/(p\beta_1-1)}\right),$$

where $\beta_1 > 1/p$. The graph of $(\omega_1^{\text{ap}})^2 Ml/c$ versus β_1 is given in Figure 3.15, and, as predicted, it takes its maximum value of $2p/(2+p)$ at $\beta_1 = 2/p$.

We note that for any admissible β_1, β_2

$$\frac{c}{Ml}\frac{p}{1+p} \leq (\omega_1^{\text{ap}})^2 \leq \frac{c}{Ml}\frac{2p}{2+p}, \tag{3.80}$$

which delivers both the upper and the lower bounds for the first eigenfrequency of the two-legged resonator. This is also illustrated by the diagram in Figure 3.15, where the limit $\beta_1 \to \infty$ corresponds to the case when the leg 1 is

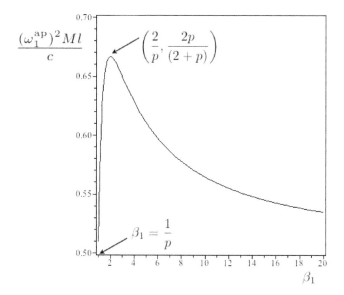

FIGURE 3.15: The lowest normalised eigenfrequency for a two-legged resonator as a function of β_1 for the case $p = 1$. In the limit $\beta_1 \to \infty$ (one leg totally damaged) the eigenfrequency approaches 0.5 which is the same as its value at $\beta_1 = 1/p$.

broken, whereas the limit $\beta_1 \to 1/p$ corresponds to the situation when the leg 2 is broken (i.e. $\beta_2 \to \infty$).

We note that the estimate (3.80) can be generalised to the case of an N-legged resonator as follows

$$\frac{c}{Ml}\frac{p}{1+p} \leq (\omega_1^{\text{ap}})^2 \leq \frac{c}{Ml}\frac{Np}{N+p}. \tag{3.81}$$

Namely, we have shown that there is only one interior point of extremum, defined by (3.79). The variable β_N can be eliminated by the constraint (3.78). Then the infimum of $(\omega_1^{\text{ap}})^2$ across the admissible values of $(\beta_1, \ldots, \beta_{N-1})$ can be achieved either at infinity (which means that all legs but one are broken and the limit for $(\omega_1^{\text{ap}})^2$ gives $\frac{c}{Ml}\frac{p}{1+p}$) or along the curve $\sum_{k=1}^{N-1} \beta_k^{-1} = p$. In turn, it follows by induction that along such a curve $(\omega_1^{\text{ap}})^2$ is greater than or equal to $\frac{c}{Ml}\frac{p}{1+p}$, as required. It is also noted that, for a fixed value of the constant p, the upper bound in (3.81) increases monotonically with increase in N.

We summarise by saying that the upper bound for the first eigenfrequency corresponds to the case when the damage is distributed uniformly across all legs within the resonator, and the lower bound corresponds to the situation when all the legs but one are broken. The maximum value of $(\omega_1^{\text{ap}})^2$ increases monotonically when we increase N, provided p remains fixed.

3.3.1.2 Dispersion

We consider out-of-plane shear Bloch–Floquet waves in the periodic structure with an elementary cell Ω shown in Figure 3.14a. Using the notation $u(\mathbf{x})$ for the displacement amplitude, we set the standard spectral problem for the Bloch–Floquet fields by imposing the Helmholtz equation

$$\mu \Delta u(\mathbf{x}) + \rho \Omega^2 u(\mathbf{x}) = 0, \quad \mathbf{x} \in \Omega, \tag{3.82}$$

with homogenous traction boundary conditions on the interior boundary

$$\frac{\partial u}{\partial n}(\mathbf{x}) = 0, \quad \mathbf{x} \in \Gamma, \tag{3.83}$$

and the Bloch–Floquet quasi-periodicity conditions on the exterior boundary of the elementary cell:

$$u(\mathbf{x} + dm_1 \mathbf{e}_1 + dm_2 \mathbf{e}_2) = u(\mathbf{x}) e^{id(K_1 m_1 + K_2 m_2)}, \tag{3.84}$$

for all $\mathbf{x} \in \Omega$, and integer m_1, m_2, where $\mathbf{e}_1, \mathbf{e}_2$ are the basis unit vectors of the Cartesian coordinate system, d is the length of the elementary cell of the doubly periodic structure, and $\mathbf{K} = (K_1, K_2)^T$ is the Bloch vector.

We solve the spectral problem numerically; given the components of the Bloch vector $\mathbf{K}$, we determine the radian frequency ω. The software COMSOL

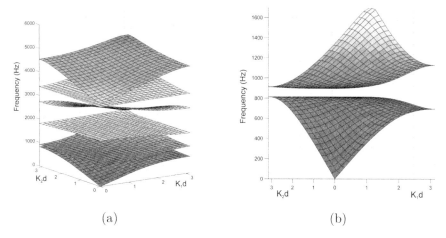

(a)   (b)

FIGURE 3.16: (a) The dispersion surfaces for Bloch–Floquet waves in a doubly periodic structure with built-in resonators. (b) The "acoustic" and "optical" dispersion surfaces for Bloch–Floquet waves, the same as in part (a), where the low frequency complete band gap is clearly visible.

Multiphysics has been used for the computations. It is assumed that **K** characterises the positions of points within the first Brillouin zone shown in Figure 3.17b. Using the notation $\omega_j(K_1, K_2)$ $(j = 1, 2, \ldots)$ for the corresponding eigenfrequencies, the surfaces $\omega = \omega_j(K_1, K_2)$ are plotted in Figure 3.16a. There is evidence of standing waves which may be expected to exist within such a periodic structure. There is a low frequency band gap between the first two dispersion surfaces. This is more clearly shown in Figure 3.16b where only the first two dispersion surfaces are shown.

As $\omega \to 0$, we can apply long-wave homogenisation and consider an effective medium as isotropic for the present case of a square elementary cell. As the frequency increases, the presence of built-in resonators becomes pronounced, and the group velocity for the waves propagating through the structured medium depends on the direction. For the low frequency regime we consider the first two dispersion surfaces, referred to as "acoustic" and "optical" surfaces, which are shown in Figure 3.16b.

(i) Dispersion diagram and low frequency modes.

Since it may be easier to work with sectional plots, we have included these in Figure 3.17a which shows dispersion curves along the segments LN, NM and LM of the Brillouin zone shown in Figure 3.17b. We compare the sectional dispersion diagram of Figure 3.17a with the dispersion diagram constructed for a frame without resonators, shown in Figure 3.17c. We can see that the

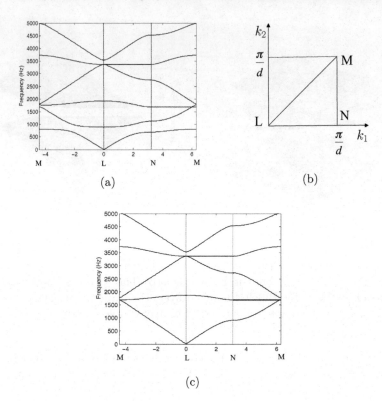

FIGURE 3.17: (a) Dispersion diagram for Bloch–Floquet waves for the out-of-plane shear problem; material is AISI steel 4340 ($\mu = 80.08$ GPa, $\rho = 7850$ kg/m^3, $d = 1$m). (b) The salient points on the first Brillouin zone. (c) Dispersion diagram for Bloch–Floquet waves for the out-of-plane shear problem for a frame without resonators.

main changes are in the low frequency range, where an additional dispersion curve appears accompanied by the low-frequency stop band.

Low frequency standing waves near the boundaries of the band gap appear to be important in a range of applications. One of the examples concerns the interaction of an incident wave with a finite width slab consisting of vertical arrays of multi-scale resonators as shown in Figure 3.20. For a specially chosen frequency of the incident wave, we observe the effect of negative refraction and consequently show that the structured interface focusses the wave produced by a localised source. Although such an effect is counterintuitive, for scalar problems of electromagnetism and finite-width slabs consisting of circular dielectric inclusions, it has been observed and discussed in [110].

Here we illustrate the influence of the multi-scale resonators, and later address a similar configuration for a full vector problem of elasticity (see Section

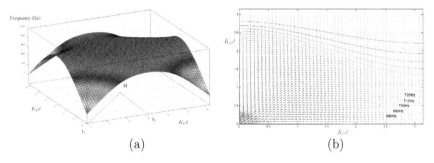

(a) (b)

FIGURE 3.18: (a) Periodic "acoustic" dispersion surface for out-of-plane shear Bloch–Floquet waves in a doubly periodic structure with built-in undamaged resonators. (b) The "acoustic" mode slowness contours for Bloch–Floquet waves which shows the direction of the group velocity (arrow plot).

3.3.3).

(ii) Group velocity as a vector function of Bloch vector components for the acoustic dispersion surface.

We plot the acoustic dispersion surface in Figure 3.18a. Although it coincides with the acoustic surface of Figure 3.16b, we choose a different orientation in order to emphasise that the points $(K_1 d, K_2 d) = (0, \pi)$ and $(K_1 d, K_2 d) = (\pi, 0)$ are saddle points, whereas $(K_1 d, K_2 d) = (\pi, \pi)$ is a maximum point for the acoustic dispersion surface extended to all admissible values of $(K_1 d, K_2 d)$. This is a common feature, noted for example in [110] (in particular Section C and Figure 7) and [36]. Compared to [110], where the periodic structure includes a square array of circular holes and the acoustic dispersion surface is symmetric with respect to $K_1 = K_2$, we do not have equivalence of the directions $K_1 = 0$ and $K_2 = 0$ since our structure includes a different type of resonator with a well-defined orientation. The effect of the directional preference is especially pronounced for the problem of vector elasticity discussed in the text below. For the current situation involving the scalar Helmholtz equation, the directional preference is visible near the edges of the Brillouin zone. In particular, Figure 3.18a for the acoustic dispersion surface shows that $\omega(\pi, 0) < \omega(0, \pi) < \omega(\pi, \pi)$, which correspond to the frequency values at the corner points of the Brillouin zone. Obviously, the group velocity is equal to zero at these stationary points, and the corresponding vibration modes represent standing waves within the periodic structure. In particular, the frequency $\omega(\pi, \pi)$ on the acoustic dispersion surface is approximated by the first eigenfrequency of the thin-legged resonator estimated in Section 3.3.1.1. We also note that $\omega(\pi, \pi)$ determines the lower bound of the first stop band on the dispersion diagram (see Figure 3.18a), and hence the position of the first stop band can be estimated analytically via the first eigenfrequency of a thin-legged

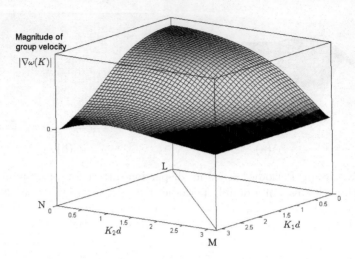

FIGURE 3.19: The magnitude of the group velocity for the acoustic dispersion surface over the first Brillouin zone.

resonator derived in Section 3.3.1.1. From Equation (3.76) with $\beta = 0$, the value of the asymptotic estimate for a steel resonator vibrating in this mode is 960 Hz. At the point L in Figure 3.17a, the first non-zero eigenfrequency is 897 Hz showing good agreement with the theoretical prediction regarding the resonator vibrating in this mode. The resonator moves in this way in the standing waves on either side of the band gap. The difference between these standing waves is in the motion of the frame. The asymptotic estimate is closer to the standing wave on the higher side of the band gap because there is less motion of the frame in this mode than in that on the lower side.

The computations, produced in Section 3.3.2 for interaction of waves with a finite-thickness structured slab, refer to a frequency close to 700 Hz, which is in the neighbourhood of the saddle point corresponding to $(K_1 d, K_2 d) = (0, \pi)$. For such a frequency, the oblique direction is "preferable" compared to the direction of the K_1-axis. The magnitude of the group velocity $|\nabla \omega(\mathbf{K})|$ at all points of the acoustic dispersion surface is shown in Figure 3.19, which indicates that largest magnitude is attained at the origin, whereas the zero values of the group velocity are linked to the corner points of the Brillouin zone. For the purpose of illustration, we also include in Figure 3.18b the slowness contours for the first dispersion surface. Five level curves between 680 Hz and 720 Hz are highlighted as these indicate the variation of the group velocity within the Brillouin zone for frequencies near the saddle point.

The implication of such a choice of frequency is the phenomenon of negative refraction in the problem of interaction between the point source of waves and a finite-thickness structured slab. From the current analysis, it follows that wave is slow in the direction of the K_1-axis compared to oblique directions. This in turn results in a phase shift of the waves arriving at the right-hand boundary of the slab. The phase difference between "secondary sources" on the right-hand boundary of the slab leads to the effect of negative refraction, as described in Section 3.3.2.

3.3.2 Filtering versus dispersion properties of out-of-plane shear Bloch–Floquet waves

In this section, we switch attention from an infinite periodic structure to a finite-thickness structured interface. Strictly, a wave generated by a point source and then interacting with the structured slab, is not a Bloch–Floquet wave. However, as discussed in [97], the analysis of Bloch–Floquet waves can be used in qualitative description of wave propagation through finite size structures. The analysis of dispersion relations for Bloch–Floquet waves proves to be particularly useful in the present case of an out-of-plane shear wave interacting with a finite-width slab, which contains an array of multi-scale resonators described in Section 3.3.1. Below, we consider and compare the cases of slabs where resonators are damaged and undamaged.

3.3.2.1 Undamaged interface

Here we consider an illustration of an interaction between a finite-width structured slab, containing built-in undamaged resonators, and a wave generated by a localised source placed at a finite distance from the interface boundary. The configuration used in the computational model is shown in Figure 3.20a.

We also adopt PML (perfectly matched layers) at the exterior boundary, apart from the slab, to suppress reflection from the boundary of the computational window. In Figure 3.20, these are shown as layers adjacent to the domains and are tuned accordingly to reduce reflections from the boundary.

The frequency chosen here is 720 Hz which is close to the lower edge of the band gap between the acoustic and optical dispersion surfaces. For the horizontal direction, across the slab, the group velocity of the waves is small, which is consistent with the computation produced in Figure 3.17a for Bloch–Floquet waves in an infinite structure. On the other hand, the group velocity in the oblique direction is positive, which enables the waves to propagate, as seen in Figure 3.20a. This anisotropy results in the phase shift between the waves reaching the other side of the structured slab. The result is the negative refraction and focussing of the image of the source clearly seen on the right-hand side from the structured slab.

3.3.2.2 Damaged interface

A similar simulation (Figure 3.20b) has been performed for an interface with all resonators possessing damaged legs, with transverse cracks, namely with the zone connecting legs and frame (see Figure 3.14a) equal to 5% of the total width of each leg. Such a change of geometry results in a small change of

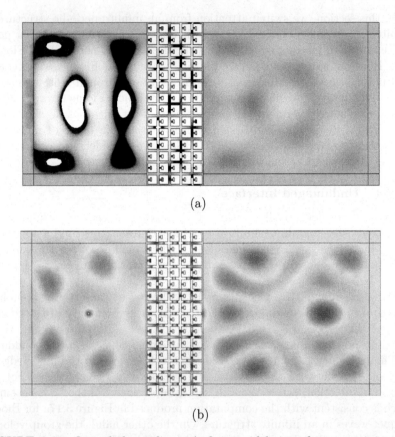

FIGURE 3.20: Out-of-plane shear. A finite-width interface containing resonators, interacting with a wave generated by a localised source: (a) undamaged resonators, with source vibrating at 720 Hz; (b) damaged resonators, with source vibrating at 694 Hz.

dispersion curves at low frequencies. Otherwise, the dispersion diagram for the corresponding infinite structure is very similar to that for the undamaged case, Figure 3.17a, with the change at low frequencies being of order 20–30 Hz. As a result, a focussing effect, although not identical to Figure 3.20a, is still present at a slightly reduced frequency of 694 Hz. This finding of very little change when the resonators are damaged, could be considered surprising. However, it can be explained by recalling that, for the out-of-plane problem, also discussed in [188], the damaged junction performs similarly to an undamaged junction unless the size of the uncracked ligament is exponentially small compared to the thickness of the thin leg of the resonator.

3.3.3 Plain strain vector problem

We now consider the case of plain strain in the same infinite, periodic elastic structure, whose elementary cell is shown in Figure 3.14a. We will concentrate on resonators with three legs only. The elastic displacement is a time-harmonic vector function, whose amplitude $\mathbf{u}(\mathbf{x})$ satisfies the following equations:

$$\mu\Delta\mathbf{u}(\mathbf{x}) + (\lambda+\mu)\nabla\nabla\cdot\mathbf{u}(\mathbf{x}) + \rho\omega^2\mathbf{u}(\mathbf{x}) = \mathbf{0}, \quad \mathbf{x} = (x_1, x_2) \in \Omega, \quad (3.85)$$

$$\boldsymbol{\sigma}^{(n)}(\mathbf{x}) = \mathbf{0}, \quad \mathbf{x} \in \Gamma, \quad (3.86)$$

where $\mathbf{u} = (u_1, u_2)^T$, and $\boldsymbol{\sigma}^{(n)}$ is the vector of tractions with entries $\sigma_j^{(n)} = \sum_k \sigma_{jk} n_k$, with σ_{jk} being the components of the Cauchy stress tensor and $\mathbf{n} = (n_1, n_2)^T$ being the outward unit normal on the interior boundary Γ.

The Bloch–Floquet quasi-periodicity conditions are also imposed on the vector function $\mathbf{u}(\mathbf{x})$ as follows:

$$\mathbf{u}(\mathbf{x} + dm_1\mathbf{e}_1 + dm_2\mathbf{e}_2) = \mathbf{u}(\mathbf{x})e^{id(K_1 m_1 + K_2 m_2)}, \quad (3.87)$$

for all $\mathbf{x} \in \Omega$, and integer m_1, m_2,

similarly to those imposed in (3.84).

3.3.3.1 Asymptotic approximations for the fundamental translational and rotational modes

Two fundamental eigenmodes are similar to those shown in Figure 3.2. In Figure 3.2a, a translational eigenmode is shown corresponding to a standing wave within the periodic structure such that the subdomain $\Omega_1 = \{(x_1, x_2) : l < x_1 < l + a, |x_2| < b/2\}$ moves in the x_2 direction, with a low frequency. The other interesting vibration mode is a rotational mode, shown in Figure 3.2b, also corresponding to a standing wave within the periodic structure. For the case when the junction between the leg and the frame is undamaged, asymptotic estimates of the frequencies of these two modes have been given in [82]. These estimates were based on the observation that the subdomains Ω_1

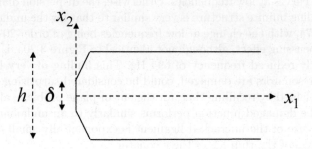

FIGURE 3.21: The geometry of the damaged junction.

and Ω_2 move like rigid solids (see Figure 3.2). Further, the legs were assumed to be thin and their inertia was neglected. The details are given in the earlier Section 3.1.3.

For the case when the junction of the leg and frame is damaged by two edge cracks perpendicular to the main axis of the leg (see Figure 3.21), effective "spring-like" junction conditions have been derived for the cases of longitudinal displacement loading [188] and bending, as discussed in Section 3.2.1 and in [69, 153].

For the case of longitudinal loading, the effective boundary condition is

$$u_1^{(0)}(0) - k_l u_1^{(0)\prime}(0) = 0, \quad k_l = \frac{4h}{\pi} \ln\left(\frac{\delta}{h}\right), \qquad (3.88)$$

where $u_1^{(0)}$ is the first order approximation to u_1, the longitudinal displacement of a leg. The leg thickness is h and the ligament length is δ. This is valid for situations of large damage with a small remaining ligament, i.e. small values of δ/h. The relation (3.88) also defines the first-order approximation of the effective compliance k_l of the junction region incorporating two cracks and a thin bridge. We note that in this first-order approximation the "spring effect" becomes significant when the thickness of the bridge is exponentially small. This implies that for a finite size bridge the second term in the left-hand side of (3.88) is small, and this relation can be replaced by the homogeneous boundary condition of the Dirichlet type.

For the case of bending, the model in [69] is accurate for large damage situations and that in [153] is accurate for small damage with a large remaining ligament. This is discussed in Section 3.2.2, and in [69], and it is summarised as

$$u_2^{(0)\prime}(0) - k_b u_2^{(0)\prime\prime}(0) = 0, \qquad (3.89)$$

where $u_2^{(0)}$ is the first order approximation to u_2, the transverse displacement of a leg.

The "spring constant" k_b is given by

$$k_b(\delta) = \begin{cases} \frac{(5-2\nu)h^3}{6\pi(1-\nu)\delta^2}, & \frac{\delta}{h} \leq 0.4, \\ \frac{3}{2}\pi h \left(1 - \frac{\delta}{h}\right)^2 q\left(1 - \frac{\delta}{h}\right), & \frac{\delta}{h} > 0.4, \end{cases} \quad (3.90)$$

where the function $q(\alpha)$ is

$$q(\alpha) = 0.5335 - 0.929\alpha + 3.500\alpha^2 - 3.181\alpha^3 + 5.793\alpha^4, \quad (3.91)$$

and ν is the Poisson's ratio of a leg.

These conditions may be used for the derivations of the translational and rotational eigenfrequencies in the case of damaged junctions. For the translational mode, the legs are predominantly loaded in bending. The derivation follows closely that given in Section 3.1.3 for the undamaged resonator and for the out-of-plane shear situation derived in Section 3.3.1. However, the undamaged junction boundary condition $u_2^{(0)''}(0) = 0$ is replaced by the boundary condition (3.89) and this leads to an asymptotic estimate for the translational (ω_t) mode radian frequency as:

$$\omega_t = \sqrt{\frac{3E}{(1-\nu^2)}\left(\frac{h}{l}\right)^3 \left(\frac{l+k_b}{l+4k_b}\right) \left(\frac{1}{\rho_1 ab} + \frac{1}{\rho_2 \mathcal{A}(\Omega_2)}\right)}, \quad (3.92)$$

where ρ_1 and ρ_2 are the densities of Ω_1 and Ω_2, respectively, while $\mathcal{A}(\Omega_2)$ corresponds to the area of Ω_2.

For the rotational mode, the rotation of Ω_1 is caused predominantly by longitudinal motion of the legs (see Section 3.1.3). The derivation for the damaged junction in this case follows closely Section 3.1.3 except that the boundary condition (3.88) is used. This leads to an asymptotic estimate for the rotational (ω_r) mode radian frequency as:

$$\omega_r = \sqrt{\frac{24Ed^2h}{\rho_1 ab(a^2+b^2)(l+k_l)}}, \quad (3.93)$$

where $2d$ is the distance between the two outer legs.

The approximations (3.92) and (3.93) have been compared with finite element simulations using the COMSOL Multiphysics® finite element package. A frequency response analysis has been carried out for a cell of unit extent and arm width of 0.1 m containing the resonator (as in Figure 3.14a) in the case of them both being made of AISI steel 4340 ($E = 205$ GPa, $\nu = 0.28$, $\rho_1 = \rho_2 = 7850$ kg/m³). The other parameter values are $l = 0.2$ m, $a = 0.14$ m, $b = 0.3$ m, $d = 0.06$ m. The comparisons are shown for the translational mode in Figure 3.22a where the frequency is plotted as a function of damage measured by the variable δ/h. Two sets of comparisons are shown for $h = 0.01$ m and $h = 0.005$ m representing width to length values of the legs as 1:20 and 1:40 respectively.

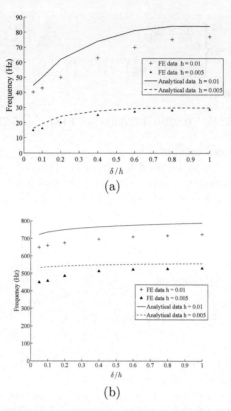

FIGURE 3.22: Comparison of the finite element results and the asymptotic approximation for the fundamental modes: (a) the low-frequency translational eigenmode; (b) the higher frequency rotational eigenmode.

It can be seen that the degree of agreement between the asymptotic approximation and the finite element computation is better for smaller values of h/l, as is expected. Corresponding results for the rotational mode are shown in Figure 3.22b.

3.3.3.2 Dispersion diagram

The dispersion diagram, showing the first six eigenfrequencies of the undamaged periodic structure discussed above, is shown in Figure 3.23 together with the path in the first Brillouin zone. There are clear standing waves at approximately 70 Hz, 720 Hz and 1270 Hz and these are significant features for all values of (K_1, K_2) in the first Brillouin zone. The first two standing waves correspond to the fundamental translational and rotational modes respectively, as discussed earlier, and the third represents large relative motion of the resonator legs. In fact, the fifth and sixth modes are close to the 1270

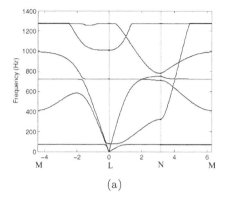
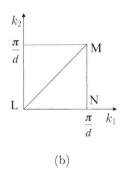

FIGURE 3.23: (a) The dispersion diagram for the vector problem of plane strain ($d = 1$); (b) the representative contour in the first Brillouin zone.

Hz frequency near to the point "M" in the Brillouin zone and the seventh mode, which is not shown, is also close to these, having an eigenfrequency of 1295 Hz at the point "M". The clustering of the fifth, sixth and seventh modes represents various large relative motions of the resonator legs.

3.3.4 Applications of multi-scale resonators in filtering and localisation of vibrations

Here we consider a type of application where dispersion properties of Bloch–Floquet waves are used to design elastic structures which filter waves within a certain frequency range. The examples in the text below involve a structured interface of finite thickness which possesses negative refraction and hence can be used for "imaging" of individual sources of waves.

Let elastic waves interact with an elastic slab (also referred to as the structured interface) shown in Figures 3.24 and 3.25. The incident wave is generated by a single point source or periodic array of point sources, in the vertical direction, placed close to the boundary of the elastic slab which is of infinite extent in the vertical direction. The structure of the slab is consistent with the models of Section 3.3.2. The displacement field is time-harmonic, and the boundary conditions on the left-hand side and on the right-hand side of the domain are modelled by perfectly matched layers.

Firstly, we consider the case when periodicity conditions are set on the upper and lower surfaces of the computational domain. The "source" is modelled by a small rigid, square inclusion being moved harmonically in the vertical direction, i.e. parallel to the structured interface. In Figures 3.24 and 3.25, we show the modulus of the displacement amplitude for forced frequencies of the source of 660 Hz and 999 Hz, respectively. At this stage, we refer to the dispersion diagrams (see Figure 3.23) constructed earlier for the elementary cell

FIGURE 3.24: The structured interface with forcing frequency 660 Hz for a periodic array of sources. Bloch–Floquet conditions are set on the upper and lower horizontal boundaries, and an elementary cell of the vertical periodic array is shown here. A periodic pattern is seen on the image side, on the right from the interface layer.

with a thin-legged resonator. Although the transmission problem is formally different from the Bloch–Floquet spectral problem, we can use the dispersion diagram for prediction of the reflection/transmission pattern. Namely, for frequencies within any stop bands, the interface will reflect a substantial part of the energy of the incident wave. Here, we draw attention to the branch with negative group velocity (see Figure 3.23a) within the range of frequencies (800 Hz, 1250 Hz). It is suggested that within this range the wave pattern may lead to the effect of negative refraction.

The image in Figure 3.24 shows predominantly plane wave fronts with no obvious evidence of focussing. At this frequency of 600 Hz, there is no evidence of negative group velocity displayed on the dispersion diagram (Figure 3.23) in the horizontal (LN) direction. However, the focussing effect is visible in Figure 3.25 where the frequency of the source oscillation is within the (800 Hz, 1250 Hz) range.

The next numerical experiment involves a single source, rather than a periodic array of sources, interacting with the structured interface. The computational domain is shown in Figures 3.26 and 3.27. The periodicity conditions are maintained on the upper and lower boundaries of the structured interface. On the remaining parts of the horizontal boundaries, the periodicity conditions are replaced by the perfectly matched layers absorbing the outgoing waves and hence removing reflection from the boundary of the computational domain. In the physical configuration this corresponds to a single source placed on the left from the infinitely long, structured interface. The source is moving harmonically in time in the vertical direction, i.e. parallel to the structured

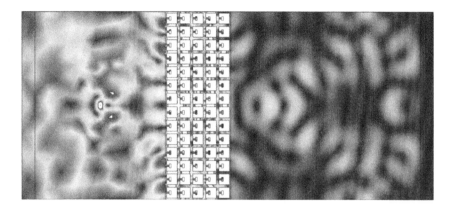

FIGURE 3.25: The structured interface with forcing frequency 999 Hz for a periodic array of sources. Imaging by a flat interface is apparent. Bloch–Floquet conditions are set on the upper and lower horizontal boundaries, and an elementary cell of the vertical periodic array is shown here.

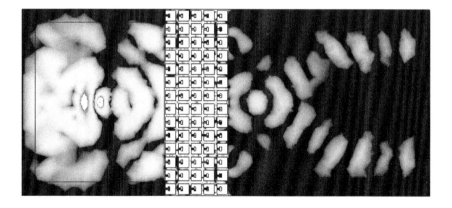

FIGURE 3.26: The computation for a single source interacting with the structured interface. A rigid inclusion is moving vertically on the left side from the interface layer. At a frequency of 950 Hz, the image of the source is clearly visible on the right from the interface layer.

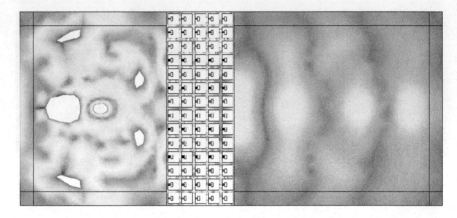

FIGURE 3.27: The source of vibrations is created by a rigid inclusion moving horizontally. At the frequency of 800 Hz, a narrow beam has been created on the right side from the interface layer.

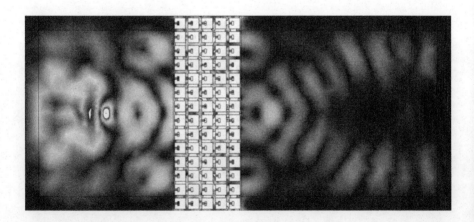

FIGURE 3.28: The same configuration as in Figure 3.26 at a frequency of 945Hz with the left-hand vertical layer being replaced by damaged resonators. The damage is introduced in the foundations of the resonator legs.

interface. Choosing a forcing frequency of 950 Hz, within the interval corresponding to a branch with negative group velocity on the dispersion diagram, we expect to see the focussing effect. This is confirmed by the computation, where the image of the source is clearly visible on the right side from the structured interface in Figure 3.26.

For this particular orientation of thin-legged resonators within the interface, a simple explanation of the focussing effect can be given. At the selected frequency of 950 Hz, according to the dispersion diagram in Figure 3.23, the group velocity for the wave propagating across the layer in the horizontal direction is smaller compared to the waves propagating in the oblique directions. Hence, this induces a phase shift for waves initiated at the right-hand boundary. Assuming that every junction region within the interface can be considered as a source of waves emanating into the right-hand medium and coupling this with the phase shift we arrive at plane waves intersecting with each other, as shown in Figure 3.26, and also in Figure 3.25 corresponding to a periodic array of sources. This replicates the effect of focussing. This effect is frequency dependent and, as the frequency is reduced to 800 Hz, we observe a narrow beam being created on the image side as shown in Figure 3.27.

The presence of resonators with damaged legs can lower the frequency at which focussing occurs. This is illustrated in Figure 3.28 where the left-hand first vertical layer of resonators has been damaged. All the legs of all of these resonators in the layer have been equally damaged at the foundation with $\delta/h = 0.05$ (Figure 3.21). The configuration is otherwise the same as in Figure 3.26. The frequency at which focussing takes place has been reduced from 950 Hz to 945 Hz.

We have demonstrated a non-trivial connection between three classes of models involving multi-scale resonators: (a) a model for a single multi-legged resonator with damage or otherwise; (b) a model of Bloch–Floquet waves interacting with a doubly periodic structure whose elementary cell contains a multi-scale resonator; (c) a dynamic response of a finite-width structured interface with built-in multi-scale resonators.

The first model allows for an analytical asymptotic treatment which leads to an estimate of frequencies corresponding to a low frequency stop band for Bloch–Floquet waves in a doubly periodic structure. Furthermore, the frequency response analysis for a single elementary cell yields dispersion surfaces and accurate information about the group velocities.

Problems of optimal design for finite-size micro-structures can be addressed by treating the structures as waveguides and hence using the properties of Bloch–Floquet waves analysed in the class (b) of the models mentioned. Illustrations include interaction of waves with a finite-width structured interface. The negative refraction is one of the interesting outcomes of the model, which has been fully explained in classical terms within the framework of the proposed approach.

Chapter 4

Dynamic response of elastic lattices and discretised elastic membranes

For waves in periodic structures, dispersion is a very important feature, which can be analysed using dispersion diagrams representing frequency as a function of the Bloch vector, as illustrated in the previous chapters of the present book. In particular, stop bands identify intervals of frequency where waves are exponentially localised.

In the present chapter we discuss a selection of problems where dynamic localisation occurs in periodic two-dimensional lattices.

It is a fascinating feature of periodic media that a lattice, which is isotropic in the long-wavelength limit, may exhibit a strong anisotropy at higher frequencies, as discussed in [49]. Localised waveforms in scalar lattice systems were previously studied in [8, 92, 93, 151]. This chapter is based on the results presented in the papers [49, 50, 131].

4.1 Stop-band dynamic Green's functions and exponential localisation

The one-dimensional example considered Section 2.5.4 illustrates localisation for waves within a uniform one-dimensional chain of masses placed on an elastic foundation and connected by massless springs. Two stop bands have been identified, where localised waveforms occur. One such stop band occurs at low frequencies, which has been achieved by setting a spring-like support for each nodal point of the lattice. The second stop band is semi-infinite, and is a common feature of spring-mass systems, where the inertia of connecting springs is neglected.

In the one-dimensional case considered earlier in Section 2.2 (cf. Equation (2.17)), the governing equations for the nodal displacements u_n are

$$\mu(u_{n+1} + u_{n-1} - 2u_n) + m\omega^2 u_n = 0, \ n \text{ is integer},$$

where μ is an elastic stiffness of the lattice links, m is the point mass placed at each nodal point, and ω is the radian frequency. This example has also provided a constructive method to design an exponentially localised waveform by a perturbation of mass at a given nodal point. In particular, the formulae (2.66) and (2.68) give the perturbation of mass required in order to create localised waveforms of a given frequency.

Here, we discuss a two-dimensional example of a discretised membrane, represented by a uniform lattice, with emphasis on dynamic anisotropy in the regime of high-frequency localised vibrations. The governing equations for the time-harmonic nodal out-of-plane displacements $u_{n,l}$ take the form

$$\mu(u_{n+1,l}+u_{n-1,l}+u_{n,l+1}+u_{n,l-1}-4u_{n,l})+m\omega^2 u_{n,l} = 0, \quad n,l \text{ are integer.} \quad (4.1)$$

We note that, in contrast to the case when an elastic foundation is present, there exists no low-frequency stop band for the present system.

4.1.1 Localised Green's function for the square lattice

The paper [107] presents an analysis of dynamic Green's functions for square lattices in regimes which enable propagating sinusoidal waves. These solutions correspond to a point source producing outgoing waves, which satisfy the standard radiation conditions at infinity.

In contrast, [131] deals with the so-called stop band Green's functions for frequencies of vibrations which are outside the admissible pass band, resulting in exponentially localised waves. We use the notation $\Omega = \omega^2$, and assume that the frequency is sufficiently high, above the pass band corresponding to sinusoidal waves. Subject to the normalisation, used here, the spectral parameter Ω is chosen in such a way that $\Omega > 8$.

The discrete Fourier transform is applied to time-harmonic waves in the square lattice governed by Equation (4.1). Using the notation u^{FF} for the double Fourier transform of the displacement field corresponding to a time-harmonic point source, of unit magnitude, placed at the origin in the central cell of the lattice we deduce that

$$u^{FF}(k,q) = \frac{1}{2(1-\cos k) + 2(1-\cos q) - \Omega}. \quad (4.2)$$

After the Fourier inversion, we obtain the integral representation for the nodal displacements

$$u(m,n) = \frac{1}{4\pi^2} \int_{-\pi}^{\pi} \int_{-\pi}^{\pi} \frac{\exp[-i(km+qn)]}{2(1-\cos k) + 2(1-\cos q) - \Omega} \, dk \, dq. \quad (4.3)$$

We note that the Fourier transform with respect to m has the form

$$u^F(k,n) = \frac{1}{\pi} \int_0^{\pi} \frac{\cos(qn)\,dq}{2(1-\cos k) + 2(1-\cos q) - \Omega}$$

$$= \frac{(-1)^{n+1}}{2\sqrt{a^2-1}} \left(a - \sqrt{a^2-1}\right)^{|n|}, \quad (4.4)$$

where $\Omega > 8$, and
$$a = \frac{\Omega}{2} - 2 + \cos k, \quad a > 1.$$
Thus, the nodal displacements corresponding to the exponentially localised waveform of frequency $\omega = \sqrt{\Omega}$ are
$$u(m,n) = \frac{(-1)^{n+1}}{\pi} \int_0^\pi \frac{1}{2\sqrt{a^2-1}} \left(a - \sqrt{a^2-1}\right)^{|n|} \cos(km)\, dk. \quad (4.5)$$

In the central cell, at the origin, we have $m = n = 0$. When $\Omega \gg 1$ the displacement at the central cell has the asymptotic approximation:
$$u(0,0) \sim -\frac{1}{2\sqrt{(\Omega/2 - 2)^2 - 1}}. \quad (4.6)$$

In particular, when $\Omega > 8$, the asymptotic approximation (4.6) is sufficiently close to the exact result (4.5) that it can be used for numerical evaluations.

4.1.2 Dispersion and dynamic anisotropy

The dispersion equation for Bloch–Floquet waves in the square lattice has the form
$$\omega^2 = 4 - 2\cos k_x - 2\cos k_y. \quad (4.7)$$
When $\omega > \sqrt{8}$ this equation does not have real solutions, and we use the following substitution: $k_x = \pi + iq_x$, $k_y = \pi + iq_y$. Hence
$$\Omega = \omega^2 = 4 + 2\cosh q_x + 2\cosh q_y. \quad (4.8)$$

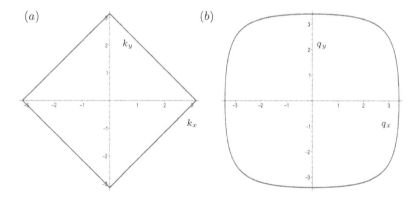

FIGURE 4.1: Dispersion contours: (a) the sinusoidal wave slowness contour, corresponding to $\Omega = 4$, and (b) the exponential wave slowness contour, which corresponds to $\Omega = 36$.

The slowness contours in Figure 4.1 illustrate two cases: (a) the case of sinusoidal waves and (b) exponential waveforms. The normal to the slowness contour indicates the direction of the group velocity, and for isotropic media the slowness contours are concentric circles centred at the origin. The shapes of the slowness contours in both cases shown in Figure 4.1 are not circular, and hence the lattice features dynamic anisotropy at the frequencies used here.

It is worthwhile mentioning that the slowness contour in Figure 4.1a, for $\Omega = 4$, is a square whose vertices lie on the coordinate axes. Hence, for values of the spectral parameter Ω close to 4, the group velocity is aligned with one of the diagonals $k_x = \pm k_y$. On the other hand, for $\Omega > \sqrt{8}$ and, for sufficiently high values of the spectral parameter, the corresponding slowness contour, in the (q_x, q_y)-plane, becomes close to a square, whose sides are aligned with the coordinate axes, as shown in Figure 4.1b.

Furthermore, when $\Omega > 8$, we can evaluate the localisation rate in the evanescent waveforms and show that it varies for different directions within the lattice.

For the square lattice, the x- and y-directions are equivalent, of course. Hence, we can consider one of these principal directions. For an exponential waveform, aligned in the y-direction, consider the upper half-plane, and with $q_x = 0$, the displacement $u(0, n)$ at the nodal points along the y-axis takes the form

$$u(0, n) = \exp(i\pi n - q_y n). \tag{4.9}$$

Equation (4.8), implies that

$$\exp(-q_y) = \lambda_{0n} = \Omega/2 - 3 - \sqrt{(\Omega/2 - 3)^2 - 1}. \tag{4.10}$$

Along the diagonal direction of the square lattice, where $q_x = q_y = q, m = n$, in the first quadrant of the plane, the nodal displacement is defined by

$$u(n, n) = \exp(2\pi i n - 2qn) = \exp(2\pi i n - 2qn) = \exp(-\sqrt{2}qr), \tag{4.11}$$

where $r = \sqrt{2}n$ is the distance to the origin. We further deduce that

$$\exp(-\sqrt{2}q) = \lambda_{nn} = \left(\Omega/4 - 1 - \sqrt{(\Omega/4 - 1)^2 - 1}\right)^{\sqrt{2}}, \tag{4.12}$$

where the quantities λ_{0n} and λ_{nn} characterise localisation of the wave in two different directions.

4.1.3 Asymptotics along the principal axes of the lattice

When $m = 0$ and $n \to \infty$, the integral (4.5) can be evaluated asymptotically. We take into account that the main contribution comes from the neighbourhood of $k = \pi$, and use the expansions

$$\cos k \sim -1 + (1/2)(\pi - k)^2, \tag{4.13}$$

$$a - \sqrt{a^2 - 1} \sim (c - \sqrt{c^2 - 1})\left[1 - \frac{(\pi - k)^2}{2\sqrt{c^2 - 1}}\right], \quad c = \frac{\Omega}{2} - 3, \quad (4.14)$$

$$\left[1 - \frac{(\pi - k)^2}{2\sqrt{c^2 - 1}}\right]^{|n|} \sim \exp\left[-\frac{(\pi - k)^2 |n|}{2\sqrt{c^2 - 1}}\right]. \quad (4.15)$$

Thus,

$$u(0, n) \sim \frac{(-1)^{n+1}}{2\sqrt{c^2 - 1}}\left(c - \sqrt{c^2 - 1}\right)^{|n|} \frac{1}{\pi}\int_{\pi - \varepsilon}^{\pi} \exp\left[-\frac{(\pi - k)^2 |n|}{2\sqrt{c^2 - 1}}\right] dk, \quad (4.16)$$

where $0 < \varepsilon \ll 1$. Further simplification gives

$$\frac{1}{\pi}\int_{\pi - \varepsilon}^{\pi} \exp\left[-\frac{(\pi - k)^2 |n|}{2\sqrt{c^2 - 1}}\right] dk = \sqrt{\frac{2\sqrt{c^2 - 1}}{|n|}} \frac{1}{\pi}\int_0^{\infty} e^{-x^2} dx$$

$$= \sqrt{\frac{\sqrt{c^2 - 1}}{2\pi |n|}}, \quad (4.17)$$

and

$$u(0, n) \sim \frac{(-1)^{n+1}}{\sqrt{8\pi\sqrt{c^2 - 1}}}\left(c - \sqrt{c^2 - 1}\right)^{|n|} \frac{1}{\sqrt{|n|}} \quad \text{as } n \to \infty. \quad (4.18)$$

4.1.4 Asymptotic approximation along the diagonal $m = n$

For the case when $m = n$, in the evaluation of the integral (4.3) it is convenient to use the substitution

$$k = x + y, \quad q = x - y \quad (4.19)$$

to deduce

$$u(m, m) = -\frac{1}{\pi^2}\int_0^{\pi}\int_0^{\pi} \frac{\cos(2mx)}{\Omega - 4 + 4\cos(x)\cos(y)} dx\, dy. \quad (4.20)$$

Furthermore, if $m \gg 1$, while integrating with respect to x, the neighbourhoods of $y = 0$ and $y = \pi$ give the leading contribution, and one can use the expansions

$$\cos y \sim 1 - \frac{y^2}{2} \quad \text{as } y \to 0,$$

$$\cos y \sim -1 + \frac{(\pi - y)^2}{2} \quad \text{as } y \to \pi.$$

Thus, we find

$$u(m,m) \sim -\frac{2}{\pi}\int_0^\infty \left[\frac{1}{\pi}\int_0^\pi \frac{\cos(2mx)}{\Omega - 4 + 4(1 - y^2/2)\cos(x)}\,dx\right]dy$$

$$\sim -\frac{1}{\sqrt{\Omega^2 - 8\Omega}}\frac{2}{\pi}\int_0^\infty \left(\frac{\sqrt{1-b^2}-1}{b}\right)^{2|m|}dy, \qquad (4.21)$$

with $b = \dfrac{4(1-y^2/2)}{\Omega - 4}$.

Furthermore, using the asymptotic representations

$$\left(\frac{\sqrt{1-b^2}-1}{b}\right)^{2|m|} \sim \left(\frac{\sqrt{1-b_0^2}-1}{b_0}\right)^{2|m|}\left(1 - \frac{y^2}{2\sqrt{1-b_0^2}}\right)^{2|m|}$$

$$\sim \left(\frac{\sqrt{1-b_0^2}-1}{b_0}\right)^{2|m|} \exp\left(-\frac{|m|y^2}{\sqrt{1-b_0^2}}\right), \quad b_0 = \frac{4}{\Omega-4}, \qquad (4.22)$$

we deduce

$$u(m,m) \sim -\frac{1}{\sqrt{\pi}\sqrt{\Omega^2-8\Omega}}\left(\frac{\sqrt{1-b_0^2}-1}{b_0}\right)^{2|m|}\sqrt{\frac{\sqrt{1-b_0^2}}{|m|}}. \qquad (4.23)$$

In terms of the distance r from the origin, this formula becomes

$$u(m,m) \sim -\frac{1}{\sqrt{\pi}\sqrt{\Omega^2-8\Omega}}\left(\frac{\sqrt{1-b_0^2}-1}{b_0}\right)^{r\sqrt{2}}\sqrt{\frac{\sqrt{2(1-b_0^2)}}{r}}, \qquad (4.24)$$

where $r = \sqrt{2}m \gg 1$.

4.1.5 Localisation exponents

We note that the quantities (4.10) and (4.12) characterise the rate of localisation in (4.18) and (4.24). Earlier, in Section 2.7.2.1 on page 44, we introduced the term *"localisation exponent"* (see [8] for more detail). In the present case, the localisation exponent takes the form

$$\lambda_{0n} = c - \sqrt{c^2 - 1} = \Omega/2 - 3 - \sqrt{(\Omega/2 - 3)^2 - 1},$$

$$\lambda_{mm} = \left(\frac{\sqrt{1-b_0^2}-1}{b_0}\right)^{\sqrt{2}} = \left(\Omega/4 - 1 - \sqrt{(\Omega/4 - 1)^2 - 1}\right)^{\sqrt{2}}. \qquad (4.25)$$

Indeed, in the asymptotic representations at infinity, the logarithmic derivative of the displacement with respect to the distance from the origin is equal to λ, where the indices have been omitted. The quantities λ_{0n} and λ_{mm} are clearly different, which represents dynamic anisotropy reflected in the different rate of localisation in different directions. We also remark that the localisation becomes stronger with the increase of the frequency, and also a stronger localisation is observed along the diagonals than along the principal directions of the square lattice.

4.2 Dynamic anisotropy and localisation near defects

In this section, we analyse the dynamic response of both square and triangular scalar lattices with emphasis on Green's functions and the diffraction patterns generated by a point load. We are also concerned with the dynamic anisotropy of discrete elastic structures in the full vector setting of planar elasticity. The analysis is focused on the directionally localised waveforms, which correspond to saddle points on the dispersion surfaces. In the present section, the term "localised" is used in a similar sense to that used in the papers [8, 151] to describe an effect where the field is predominantly confined to one or more finite width "beams" with differing orientations.

4.2.1 Primitive waveforms in scalar lattices

In this section, two different monatomic lattice geometries will be examined: square and triangular. We start by considering the general problem of the out-of-plane displacements of a regular uniform lattice in $\mathbb{R}^2$ when the lattice is harmonically forced at a single point. The assumption of uniformity is purely for convenience and can be weakened without significant additional work. Given this uniformity, we may choose the forcing point to be the origin. The lattice consists of a regular array of unit masses, connected by massless Hookean springs of unit stiffness. Let the multi-index $\boldsymbol{m} \in \mathbb{Z}^2$ label the lattice nodes such that their position is given by $\boldsymbol{x}(\boldsymbol{m}) = m_1 \boldsymbol{t}_1 + m_2 \boldsymbol{t}_2$, where $\boldsymbol{t}_1$ and $\boldsymbol{t}_2$ are the direct lattice basis vectors. We denote the set of all $\boldsymbol{q} \in \mathbb{Z}^2$ nodes connected to node $\boldsymbol{m}$ by $\mathbb{P}(\boldsymbol{m})$. The out-of-plane displacement amplitude $u_{\boldsymbol{m}}$, normalised by spring length, is then governed by the equation

$$\sum_{\boldsymbol{q} \in \mathbb{P}(\boldsymbol{m})} u_{\boldsymbol{q}} - \overline{\overline{\mathbb{P}(\boldsymbol{m})}} u_{\boldsymbol{m}} + \omega^2 u_{\boldsymbol{m}} = \delta_{\boldsymbol{m}\boldsymbol{0}}, \quad (4.26)$$

where we have omitted the factor $e^{i\omega t}$, and ω is the radian frequency. The symbol δ_{ij} is the Krönecker delta, and the symbol $\overline{\overline{\cdot}}$ denotes the cardinality (i.e. the number of nodes connected to node $\boldsymbol{m}$ in this case). We use the discrete Fourier transform

$$\mathcal{F}: u_q \mapsto U(\boldsymbol{\xi}) = \sum_{q\in(-\infty,\infty)^2} u_q e^{-i(\boldsymbol{x}(q)\cdot\boldsymbol{\xi})}, \qquad (4.27)$$

whence (4.26) may be rewritten as

$$\left(\sum_{q\in\mathbb{P}(m)} e^{i(\boldsymbol{x}(q-m)\cdot\boldsymbol{\xi})} - \overline{\mathbb{P}(m)} + \omega^2 \right) U(\boldsymbol{\xi}) = 1, \qquad (4.28)$$

Writing the parenthesised term in Equation (4.28) as $\sigma(\boldsymbol{\xi};\omega)$, the physical field can be obtained by taking the inverse Fourier transform:

$$u_m = \mathcal{F}^{-1}\left[\frac{1}{\sigma(\boldsymbol{\xi};\omega)}\right]. \qquad (4.29)$$

For the unforced problem, the transformed governing equation (4.28) reduces to $\sigma(\boldsymbol{\xi};\omega)U(\boldsymbol{\xi}) = 0$, which has the well-known solution $u(m,n) = \exp(i\boldsymbol{x}(m,n)\cdot\boldsymbol{\xi})$ where $\boldsymbol{\xi}$ satisfies $\sigma(\boldsymbol{\xi};\omega) = 0$.

4.2.1.1 Square monatomic lattice

Consider first, the uniform square lattice as illustrated in Figure 4.2 such that $\boldsymbol{t}_1 = (1,0)^{\mathrm{T}}$ and $\boldsymbol{t}_2 = (0,1)^{\mathrm{T}}$ and $\sigma(\boldsymbol{\xi};\omega) = \omega^2 - 4 + 2(\cos\xi_1 + \cos\xi_2)$. The displacement field is then given by the Fourier integral

$$u_m = \frac{1}{\|\mathcal{R}\|} \iint_{\mathcal{R}} \frac{\exp(i\boldsymbol{m}\cdot\boldsymbol{\xi})}{\sigma(\boldsymbol{\xi};\omega)} \,\mathrm{d}\boldsymbol{\xi}, \qquad (4.30)$$

where $\mathcal{R}$ is the elementary cell in the periodic reciprocal lattice. The scalar function σ is even with respect to $\boldsymbol{\xi}$ and the region $\mathcal{R}$ is symmetric about $\xi_i = 0$ ($i = 1, 2$), hence the odd terms in $\exp(i\boldsymbol{m}\cdot\boldsymbol{\xi})$ do not contribute to the integral and the Lattice Green's Function may be expressed as

$$u_m = \frac{1}{\pi^2} \iint_{[0,\pi]^2} \frac{\cos(m_1\xi_1)\cos(m_2\xi_2)}{\sigma(\boldsymbol{\xi};\omega)} \,\mathrm{d}\boldsymbol{\xi}. \qquad (4.31)$$

Alternative representations of the Lattice Green's function and detailed analysis in various frequency regimes may be found in many texts, including [49,104,107,131], in addition to later in the present monograph. For certain restrictions of ω one, but not both, of the integrals in (4.31) may be evaluated in closed form; alternatively, the integral may be converted to a semi-infinite integral over the positive semi-axis. However, for the purposes of this chapter it suffices to consider the Lattice Green's function in the form (4.31).

The dispersion surface (the zero isosurface $\{(\boldsymbol{\xi};\omega) : \sigma(\boldsymbol{\xi};\omega) = 0\}$) is shown

Dynamic response of elastic lattices and discretised elastic membranes   103

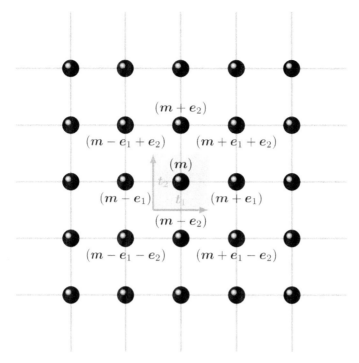

FIGURE 4.2: The monatomic square lattice and its elementary cell (shaded in grey). The lattice vectors t_1 and t_2 are also shown. The vectors e_i are defined as follows: $e_1 = (1,0)^T$ and $e_2 = (0,1)^T$.

in Figure 4.3a and has a number of interesting features. In particular, it is observed that within the Brillouin zone the surface has one maximum, attained at the four points at the corners of the Brillouin zone and four saddle points labelled $\pm A$ and $\pm B$. The saddle points all lie at the frequency $\omega = 2$, for which the slowness contour is shown in Figure 4.3b in addition to its representation on the dispersion diagram of Figure 4.3a. The saddle points lie at the vertices of the rhombic slowness contour. We also note that Figure 4.3 is fully consistent with the results presented in [8, 151].

The optical analogue of (4.31) is the so-called diffraction integral [25, 73, 145]. In optics, the term *aberration* is used to describe perturbation of the wave front away from its ideal shape as a result of a lens or diffraction grating [25, 73]. The distinction is often made between two types of aberration: *chromatic* and *monochromatic*. The latter is attributed to the geometry of the lens or grating whilst the former results from the dispersive properties of the lens. In the case of a uniform mechanical lattice, there is no such distinction since the dispersive properties arise as a result of the geometry of the medium. Hereinafter, the term *aberration* is used to describe the features of the field resulting from the dispersive properties of the lattice.

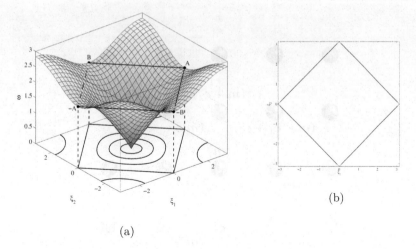

FIGURE 4.3: (a) The dispersion surface for the square cell lattice together with the projections of the level curves onto the $\omega = 0$ plane. (b) The slowness contour at the frequency coinciding with the saddle points, $\omega = 2$. The saddle points lie at the vertices of the rhombic slowness contour.

The displacement field for the square cell lattice when the forcing frequency coincides with the frequency of the saddle points $\omega = 2$ is shown in Figure 4.4. The field is determined by computing (4.31) numerically using the Gauss–Kronrod quadrature algorithm in MATLAB® for each m in a given range. The displacement field is consistent with the *star-shaped contours* studied in [8, 92, 93, 151]. However, a novel feature is observed now: the rhombic aberration in the vicinity of the point source. The effect is sensitive to perturbations in the frequency around the saddle points $\pm A$ and $\pm B$ in Figure 4.3a. For example, changing the frequency by as little as 0.01 significantly alters the diffraction pattern shown in Figure 4.4. This sensitivity can be understood in terms of the group velocity which varies rapidly in the vicinity of the saddle points. Moreover, the phenomena of *star-shaped contours* and *aberrations* is closely linked with the nature of the slowness contours. In particular, consider the slowness contour in Figure 4.3b. It is observed that the direction (but not the magnitude) is piecewise constant over the Brillouin zone. These constant directions, corresponding to the normals of the sides of the rhombus, are precisely those of the four rays shown in Figure 4.4. The group speed is maximal at the centre of each side of the rhombus and is zero at the vertices.

4.2.1.2 Stationary point of a different kind

Consider the dispersion equation $\sigma(\omega, \boldsymbol{\xi}) = \omega^2 - 4 + 2(\cos\xi_1 + \cos\xi_2) = 0$. Since $|\cos\xi_1 + \cos\xi_2| \le 2$, for any real ξ_1, ξ_2, there exist no real solutions for $\omega^2 > 8$; hence the square lattice possesses a semi-infinite stop band for

Dynamic response of elastic lattices and discretised elastic membranes 105

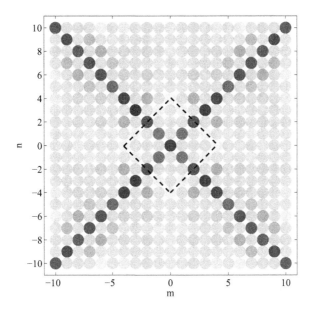

FIGURE 4.4: The out-of-plane displacement field of the square cell lattice for a forcing frequency of $\omega = 2$.

frequencies $\omega^2 > 8$ where no propagating solutions exist. For the case of free oscillations, Martin [107] found that there exist solutions of the form $u_m = (-1)^{m_1+m_2}$. These so-called *lattice waves* exist at the *resonant frequency* $\omega = 2\sqrt{2}$ which demarcates the pass band and the stop band, i.e. these are the maxima in Figure 4.3a. A similar phenomenon is observed in the case of forced excitation. In particular, Figure 4.5 shows a plot of the field for such a resonant frequency. The white (black) nodes indicate maximal positive (negative) displacement. In direct analogy to the lattice waves described in [107], the displacement of the nodes can be approximately described by $u(\bm{m}) \approx (-1)^{m_1+m_2} u(\bm{0})$. Here, no preferential direction of propagation is observed.

4.2.1.3 Triangular cell lattice

As a further example, the triangular lattice with basis vectors $\bm{t}_1 = (1, 0)^\mathrm{T}$ and $\bm{t}_2 = (1/2, \sqrt{3}/2)^\mathrm{T}$ as illustrated in Figure 4.6 is considered. In this case, the physical field has the representation

$$u_{\bm{m}} = \frac{\sqrt{3}}{4\pi^2} \iint_{\mathcal{R}} \cos[(m_1 + m_2/2)\xi_1] \cos[n\sqrt{3}\xi_2/2] \sigma^{-1}(\bm{\xi}; \omega) \, \mathrm{d}\bm{\xi}, \qquad (4.32)$$

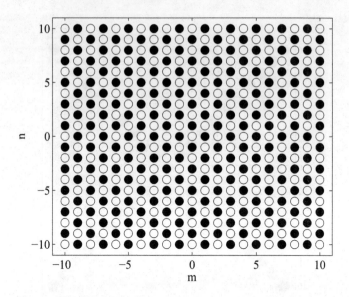

FIGURE 4.5: *Lattice waves* where the origin of the lattice is forced at the resonant frequency $\omega = 2\sqrt{2}$. White nodes indicate maximal positive displacement, while black nodes correspond to maximal negative displacement.

where $\sigma(\boldsymbol{\xi};\omega) = \omega^2 - 6 + 2\cos\xi_1 + 4\cos(\xi_1/2)\cos(\sqrt{3}\xi_2/2)$ and $\mathcal{R} = [0, 2\pi] \times [0, 2\pi/\sqrt{3}]$. The dispersion surface, together with the slowness contour for the frequency $\omega = 2\sqrt{2}$ corresponding to the position of the saddle points ($\pm A$, $\pm B$ and $\pm C$), is shown in Figure 4.7. As in the case of the square lattice, the direction of the group velocity is piecewise constant along the sides of the hexagon, with the saddle points located at the vertices. Here, six preferential directions of propagation (directions of maximal group velocity) corresponding to the perpendicular bisectors of the six sides are clearly identifiable. The displacement field for the triangular lattice when the forcing frequency is $\omega = 2\sqrt{2}$ is shown in Figure 4.8. As expected, the star-shaped waveforms with the six rays corresponding to the six discrete directions of group velocity, as indicated by the slowness contour in Figure 4.7b, are evident. Figure 4.8 is consistent with the star-shaped contours shown in [8, 151].

The determination of the position of the semi-infinite stop band requires a little more attention than in the case of a square lattice. With reference to the dispersion equation $\omega^2 - 6 + 2\cos\xi_1 + 4\cos(\xi_1/2)\cos(\sqrt{3}\xi_2/2) = 0$, the band edge corresponds to the global minimum of the function $f(\boldsymbol{\xi}) = 2\cos\xi_1 + 4\cos(\xi_1/2)\cos(\sqrt{3}\xi_2/2)$. Since the dispersion equation is periodic with respect to the elementary cell of the reciprocal lattice, $\boldsymbol{\xi}$ may be restricted to the parallelogram spanned by the two primitive vectors $\boldsymbol{b}_1 = \pi(2, -2/\sqrt{3})^{\mathrm{T}}$ and $\boldsymbol{b}_2 = (0, 4\pi/\sqrt{3})^{\mathrm{T}}$ in the reciprocal lattice. The first partial derivatives,

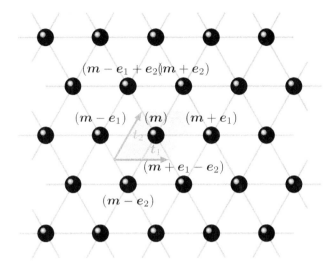

FIGURE 4.6: The monatomic triangular lattice and its elementary cell (shaded in grey). The lattice vectors t_1 and t_2 are also indicated. The vectors e_i are defined thus: $e_1 = (1,0)^T$ and $e_2 = (0,1)^T$.

Hessian determinant, and second partial derivative with respect to ξ_1 are then

$$\nabla f(\boldsymbol{\xi}) = -2 \begin{pmatrix} \sin \xi_1 + \cos(\sqrt{3}\xi_2/2) \sin(\xi_1/2) \\ \sqrt{3} \cos(\xi_1/2) \sin(\sqrt{3}\xi_2/2) \end{pmatrix}, \quad (4.33\text{a})$$

$$H(\boldsymbol{\xi}) = \frac{3}{2} \left\{ \cos \xi_1 + \cos(\sqrt{3}\xi_2) + 2 \left[\cos\left(\frac{\xi_1}{2}\right) + \cos\left(\frac{\sqrt{3}\xi_2}{2}\right) \right] \right\}, \quad (4.33\text{b})$$

and

$$\frac{\partial^2 f}{\partial \xi_1^2} = -2 \cos \xi_1 - \cos\left(\frac{\xi_1}{2}\right) \cos\left(\frac{\sqrt{3}\xi_2}{2}\right). \quad (4.33\text{c})$$

Within the irreducible Brillouin zone, the function $f(\boldsymbol{\xi})$ has stationary points at the following positions

$$\Xi = \{(0,0)^T, (\pi, \pi/\sqrt{3})^T, (0, 4\pi/\sqrt{3})^T\}. \quad (4.34)$$

Analysis of the signs of the Hessian determinant and second derivatives at the stationary points reveals that the first is a local maximum of $f(\boldsymbol{\xi})$, the second is a saddle point, whilst the third is a local minimum. Indeed, since these are the only stationary points in the irreducible Brillouin zone of the reciprocal lattice, the local extrema are global extrema. Thus, the maximum value of ω which corresponds to the minima of $f(\boldsymbol{\xi})$ is $\omega = 3$. Hence, there exists a semi-infinite stop band for frequencies $\omega > 3$, whilst propagating solutions are supported for $0 < \omega \leq 3$. The saddle point frequency, corresponding to the saddle points of $f(\boldsymbol{\xi})$, is $\omega = 2\sqrt{2}$ as stated earlier. Finally, as expected, the minimum value of ω, corresponding to the maxima of $f(\boldsymbol{\xi})$ is $\omega = 0$.

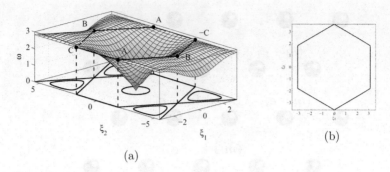

(a)

(b)

FIGURE 4.7: (a) The dispersion surface for the triangular cell lattice and (b) the slowness contour at the frequency coinciding with the saddle points, $\omega = 2\sqrt{2}$. The saddle points lie at the vertices of the hexagon.

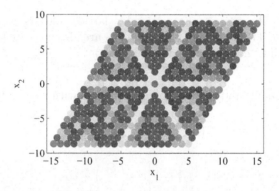

FIGURE 4.8: The magnitude of the out-of-plane displacement field of the triangular cell lattice for a forcing frequency of $\omega = 2\sqrt{2}$.

4.2.2 Diffraction in elastic lattices

This section is devoted to the analysis of the vector elasticity analogue of the problems presented in the previous section. The in-plane elasticity problem is distinct from the scalar system and presents a number of novel features and challenges. In particular, it is demonstrated that the orientation of the applied force can be used to select one or more of the preferential directions defined by the dispersive properties of the lattice. In the scalar case, the papers [8, 151] have focused on the preferential directions, *primitive waveforms*, and star-shaped contours at resonant (saddle point) frequencies. As mentioned in the previous section, these primitive waveforms and associated effects are sensitive to perturbation in the frequency around the saddle points. In contrast to the scalar problem, when working in the framework of vector elasticity it will be shown that similar star-shaped waveforms exist at frequencies other than resonant frequencies. In other words, the presence of star-shaped waveforms is not necessarily linked to the existence and position of stationary points on the dispersion surfaces.

The concept of preferential directions of propagation in discrete elastic structures has been demonstrated in [49], which built on the earlier work for the structured continuum [83] and for the discrete interface embedded within the continuum [29]. The three papers [29, 49, 83] also illustrate the effects of filtering and focussing of plane elastic waves and the formation of image points.

Consider a regular triangular array of uniform point masses arranged in $\mathbb{R}^2$ as depicted in Figure 4.6. The point masses are connected by Euler–Bernoulli beams of Young's modulus E, cross-sectional area S, second moment of inertia I, length ℓ and non-dimensional density $\varrho = \bar{\varrho}S\ell/m$. Here, the bars are used to denote dimensional quantities where a corresponding non-dimensional quantity has been defined. The links connect nodes of mass m and non-dimensional polar moment of inertia $J = \bar{J}/m\ell^2$ located at $\boldsymbol{x}(\boldsymbol{m}) = \ell(m_1 + m_2/2, m_2\sqrt{3}/2)^\mathrm{T}$. As before, we label the nodes by the multi-index $\boldsymbol{m} \in \mathbb{Z}^2$. We treat the lattice as a *frame structure*, that is, the angles at which the lattice links meet at the nodes are fixed. For harmonic waves of non-dimensional radian frequency $\omega = \bar{\omega}\ell/\sqrt{ES\ell/m}$ the equation of motion for node $\boldsymbol{m}$ may be written

$$-\mathsf{M}\omega^2 \boldsymbol{u_m} + \sum_{\boldsymbol{p} \in \mathcal{N}(\boldsymbol{m})} \mathsf{A}(\boldsymbol{p})\boldsymbol{u_{m+p}} = \boldsymbol{0}. \qquad (4.35)$$

Here $\mathsf{M} = \mathrm{diag}(1, 1, J)$ is the mass matrix, and the non-dimensional vectors $\boldsymbol{u_m} = (u_m/\ell, v_m/\ell, \theta_m)^\mathrm{T}$ denote the displacement amplitudes (two components of translation and one in-plane rotation) of node $\boldsymbol{m}$. The matrices $\mathsf{A}(\boldsymbol{p})$ describe the interaction between nodes $\boldsymbol{m}$ and $\boldsymbol{m}+\boldsymbol{p}$, i.e. the force at $\boldsymbol{m}$ as a result of motion at $\boldsymbol{m}+\boldsymbol{p}$. The set $\mathcal{N}(\boldsymbol{m})$ is the set of particles $(\boldsymbol{m}+\boldsymbol{p}, q)$ connected to node $(\boldsymbol{m}, n)$, typically this will be the set of nearest neighbours such that $\mathcal{N}(\boldsymbol{m}) = \{\boldsymbol{p} : |\boldsymbol{x}(\boldsymbol{m}+\boldsymbol{p}) - \boldsymbol{x}(\boldsymbol{m})| \leq \ell\}$, where ℓ is the bond length. It is emphasised that $\boldsymbol{0} \in \mathcal{N}(\boldsymbol{m})$, i.e. a node is connected to itself. We use the discrete Fourier transform $\mathcal{F} : \boldsymbol{u}(\boldsymbol{m}) \mapsto \boldsymbol{U}(\boldsymbol{\xi})$ (cf. Equation (4.27)) whence equation (4.35) can be re-written $\sigma(\boldsymbol{\xi}; \omega)\boldsymbol{U}(\boldsymbol{\xi}) = \boldsymbol{0}$, with the solvability criterion yielding the dispersion equation for the system

$$\det \sigma(\boldsymbol{\xi}; \omega) = 0, \qquad (4.36)$$

and the matrix $\sigma(\boldsymbol{\xi};\omega)$ is most conveniently expressed as

$$\sigma(\boldsymbol{\xi};\omega) = A_2 e^{-i\boldsymbol{x}(1,0)\cdot\boldsymbol{\xi}} + \mathsf{R}\,(1)\,A_2\mathsf{R}\,(1)^{\mathrm{T}}\, e^{-i\boldsymbol{x}(0,1)\cdot\boldsymbol{\xi}}$$
$$+ \mathsf{R}\,(2)\,A_2\mathsf{R}\,(2)^{\mathrm{T}}\, e^{-i\boldsymbol{x}(-1,1)\cdot\boldsymbol{\xi}} + \mathsf{R}\,(3)\,A_2\mathsf{R}\,(3)^{\mathrm{T}}\, e^{-i\boldsymbol{x}(-1,0)\cdot\boldsymbol{\xi}}$$
$$+ \mathsf{R}\,(4)\,A_2\mathsf{R}\,(4)^{\mathrm{T}}\, e^{-i\boldsymbol{x}(0,-1)\cdot\boldsymbol{\xi}} + \mathsf{R}\,(5)\,A_2\mathsf{R}\,(5)^{\mathrm{T}}\, e^{-i\boldsymbol{x}(1,-1)\cdot\boldsymbol{\xi}}$$
$$+ \sum_{t=0}^{5} \mathsf{R}\,(t)\,A_1\mathsf{R}\,(t)^{\mathrm{T}}. \quad (4.37\mathrm{a})$$

The matrices A_1 and A_2 have the form

$$A_1 = \begin{pmatrix} \eta\cot\eta & 0 & 0 \\ 0 & & \\ 0 & \tilde{A}_1 & \end{pmatrix},$$

$$A_2 = \begin{pmatrix} -\eta\csc\eta & 0 & 0 \\ 0 & & \\ 0 & \tilde{A}_2 & \end{pmatrix},$$

(4.37b)

with

$$\tilde{A}_1 = \begin{pmatrix} \dfrac{\beta\lambda^3(\cosh\lambda\sin\lambda - \cos\lambda\sinh\lambda)}{2(1 - \cos\lambda\cosh\lambda)} & \dfrac{\beta\lambda^2\sin\lambda\sinh\lambda}{2(1 - \cos\lambda\cosh\lambda)} \\ \dfrac{\beta\lambda^2\sin\lambda\sinh\lambda}{2(1 - \cos\lambda\cosh\lambda)} & \dfrac{\beta\lambda(\sin\lambda\cosh\lambda - \cos\lambda\sinh\lambda)}{2(1 - \cos\lambda\cosh\lambda)} \end{pmatrix},$$

$$\tilde{A}_2 = \begin{pmatrix} \dfrac{\beta\lambda^3(\sin\lambda + \sinh\lambda)}{2(\cos\lambda\cosh\lambda - 1)} & \dfrac{\beta\lambda^2(\cosh\lambda - \cos\lambda)}{2(1 - \cos\lambda\cosh\lambda)} \\ \dfrac{\beta\lambda^2(\cos\lambda - \cosh\lambda)}{2(1 - \cos\lambda\cosh\lambda)} & \dfrac{\beta\lambda(\sin\lambda - \sinh\lambda)}{2(\cos\lambda\cosh\lambda - 1)} \end{pmatrix}.$$

The natural frequencies of flexural and longitudinal waves of the lattice links are characterised by the non-dimensional parameters $\lambda^2 = \omega\sqrt{\varrho S\ell^2/I}$ and $\eta = \omega\sqrt{\varrho}$, respectively. The matrices $\mathsf{R}(t)$ are the matrices of a counter-clockwise rotation through $t\pi/3$ about the x_3-axis:

$$\mathsf{R}(t) = \begin{pmatrix} \cos(t\pi/3) & -\sin(t\pi/3) & 0 \\ \sin(t\pi/3) & \cos(t\pi/3) & 0 \\ 0 & 0 & 1 \end{pmatrix}. \quad (4.38)$$

The displacement amplitude $\boldsymbol{u_m} \in \mathbb{C}^3$ is a three-dimensional vector with the first two components corresponding to translational motion and the third describing micropolar rotations. The field has the same representation as (4.32) and the dispersion equation is $\det\boldsymbol{\sigma}(\boldsymbol{\xi};\omega) = 0$.

Dynamic response of elastic lattices and discretised elastic membranes 111

TABLE 4.1: The material and geometrical parameters used to produce the dispersion surfaces and finite element computations.

Property	Value
Young's Modulus (E)	200 GPa
Second Moment of Inertia (I)	349×10^{-8} m^4
Cross Sectional Area (S)	2.12×10^{-3} m^2
Beam Density ($\bar{\varrho}$)	7850 kg m^{-2}
Beam Length (ℓ)	1 m
Nodal Mass (m)	91.531 kg
Polar Mass Moment of Inertia ($\bar{J}$)	66.568 kg m^2

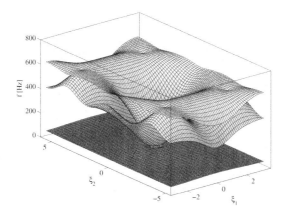

FIGURE 4.9: The first three dispersion surfaces for the triangular lattice of masses connected by Euler–Bernoulli beams in $\mathbb{R}^2$.

4.2.2.1 Dispersive properties

The dispersion surfaces for the infinite lattice system are shown in Figure 4.9. In this case, it is convenient to work with dimensional units. The material parameters used to produce the dispersion surfaces are detailed in Table 4.1. Here the focus will be on the second and third surfaces, which contain saddle points. Figure 4.10e shows the slowness contour for the resonant frequency $f = 615.8$ Hz. The contour exhibits the same characteristic hexagonal shape as the slowness contour of the scalar triangular lattice (see Figure 4.7b) suggesting that the characteristic shape of the slowness contours are a feature of the geometry of the lattice. It is emphasised that the governing equations for vector elasticity are significantly different from those of scalar problems. As in the scalar case, the six preferential directions can be identified as the normals to the edges of the hexagon, although here, the slowness

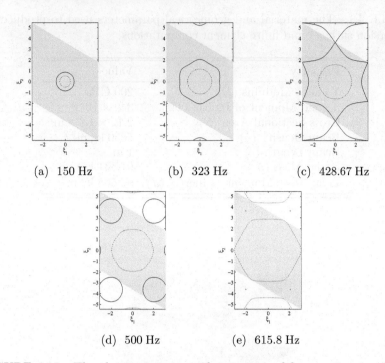

FIGURE 4.10: The slowness contours for a range of frequencies starting at 150 Hz (a) and ending at the saddle point frequency near the band edge (e). The solid lines correspond to the lower conical surface, whilst the dotted lines correspond to the upper conical surface. The elementary cell in the reciprocal space is shaded in grey.

contour is rotated by $\pi/2$ compared with the scalar case. Now consider Figure 4.10b, which shows the slowness contours for the frequency $f = 323\,\mathrm{Hz}$. Here a similar hexagonal slowness contour is observed, with the same orientation as for the scalar case. However, $f = 323\,\mathrm{Hz}$ does not correspond to a resonant frequency, that is there are no saddle points on the dispersion surfaces which coincide with $f = 323\,\mathrm{Hz}$. Nevertheless, the six preferential directions of propagation are clearly visible. This is in contrast to the scalar problems considered earlier in this chapter, in Section 4.2.1, and in the papers [8, 151], where these polygonal-like slowness contours were associated exclusively with saddle points.

4.2.2.2 Forced problem in elastic structured media

Following the structure of the previous section on scalar lattices, the forced in-plane problem is now considered. In particular, the monatomic uniform triangular lattice described above is subjected to a concentrated load (either linear or torsional) at a single lattice point. The triangular lattice is chosen because it is isotropic in the long-wavelength limit [49]. Here, the emphasis is on the dynamic anisotropy at higher frequencies and in particular on the existence of localised primitive waveforms, in direct analogy to the scalar case previously considered. The finite element software COMSOL Multiphysics® is used to determine the displacement field. The lattice nodes on the boundary are fixed and PML-like absorbing boundary conditions are applied to the lattice links in the vicinity of the boundary nodes in order to reduce reflection. The harmonic disturbance is generated in a similar fashion as in the scalar case: by prescribing a time-harmonic displacement of magnitude 10^{-6} m in a given direction or a time-harmonic rotation at node $(0,0)$. The material parameters are as detailed in Table 4.1. A selection of the displacement amplitude fields for various forcing orientations and frequencies are shown in Figure 4.11. Figures 4.11a, 4.11c and 4.11e correspond to an excitation frequency of 323 Hz for which the slowness contours are shown in Figure 4.10b. Similarly, Figures 4.11b, 4.11d and 4.11f correspond to a excitation frequency of 615.8 Hz for which the slowness contours are shown in Figure 4.10e. It is observed that the slowness contours of Figure 4.10e correspond to the saddle point on the upper dispersion surface of Figure 4.9. The other frequency of 323 Hz (see Figure 4.10b) is not a saddle point frequency. However, the slowness contour corresponding to the lower dispersion surface has six segments with almost zero curvature, and the normal vector to this slowness contour shows preferential directions at this particular frequency. The slowness contours for the saddle point at 428.67 Hz on the lower dispersion surface are shown in Figure 4.10c; the contour corresponding to the lower dispersion surface contains corner points but the curvature of the smooth parts of the boundary is large.

The displacement fields have a number of interesting features. Firstly, one may observe the so-called *primitive waveforms* already demonstrated in the scalar case [8, 92, 93, 151]. However, in contrast to the scalar case, the presence of these star-shaped contours (or localised waveforms) is not associated with the resonant frequencies as identified in [8, 151]. In the present problem, the localised waveforms are associated with frequencies where the slowness contours exhibit strong preferential directions (see Figure 4.10). For the two frequencies considered, the six preferential directions of propagation corresponding to the outward unit normals may be identified. It is emphasised that only Figures 4.10c and 4.10e correspond to a resonant frequency and yet the localised waveforms persist at the non-resonant frequency of 323 Hz due to the shape of slowness contour in Figure 4.10b. In contrast to the scalar lattice, where the applied loading was anti-planar and hence isotropic, the in-plane elasticity problem allows the freedom to choose any in-plane direction of the applied loading. As can be observed from the computations shown in Figure 4.11, the orientation of the applied force has a significant effect on the resultant field. The effect is strongest in directions that have a component of the group velocity perpendicular to the applied force. The hexagonal aberra-

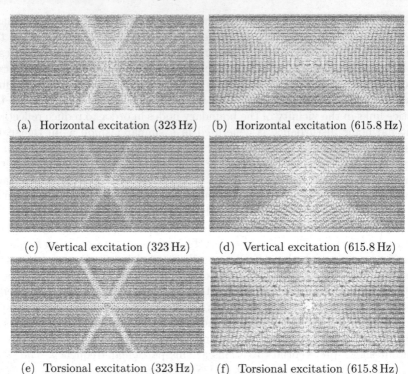

FIGURE 4.11: Finite element computations showing the magnitude of the [real] displacement amplitude fields for different types of applied load. On figures (a),(c) and (e), the excitation frequency is 323 Hz and 615.8 Hz in (b),(d) and (f). The white regions are those regions where the displacement field is outside the range.

tions and wave envelopes are evident at the resonant frequency (see Figures 4.11b, 4.11d and 4.11f).

For comparison, Figure 4.12 also shows the displacement amplitude at the saddle point frequency of 428.67 Hz, with the slowness contours shown in Figure 4.10c. Three types of loading, similar to those of Figure 4.11, are shown. Although the directional preference is clearly visible, the aberration is more pronounced for this case compared to Figure 4.11.

The different orientations of the hexagonal slowness contours and hence, the different preferential directions of propagation for different frequencies are a novel feature of the elastic lattice, which are absent in the scalar cases. One may envisage applications in shielding and focussing of elastic waves where this "switch" in preferential direction coupled with the ability to "select" a given direction via the applied force could be useful. The frequency at which this switching of preferential direction occurs is exactly the saddle point frequency of 428.67 Hz. This saddle point also marks the frequency at which a

(a) Horizontal excitation (b) Vertical excitation (c) Torsional excitation

FIGURE 4.12: Finite element computations showing the magnitude of the displacement amplitude fields for different types of the applied force. The excitation frequency is 428.67 Hz, which coincides with the saddle point frequency for which the slowness contours are shown in Figure 4.10c. The colours indicate the magnitude of the displacement field from blue (zero) to red (maximal). The white regions are those regions where the displacement field is above the range.

similar rotation in the hexagonal-like contours occurs. As can be seen from Figure 4.10c, this is also the frequency at which the slowness contours intersect at two corners and the centre of each side of the elementary cell in the reciprocal lattice.

We have examined the dynamic anisotropy of both scalar and elastic discrete systems. In particular, extending the previous work with scalar lattices, we have demonstrated the presence of directionally localised waveforms in elastic lattices which are isotropic in the long-wavelength limit. These waveforms are identified with regions on the dispersion surfaces and slowness contours with several directions of strongly preferential propagation. The presence of aberrations in the displacement fields, corresponding to the shape of the slowness contours have been observed and a connection has been made with the notion of aberration in optics.

4.3 Localisation near cracks/inclusions in a lattice

A well-known and interesting feature of discrete media is the existence of pass and stop bands, as discussed in Chapter 2. In the present section, localised defect modes associated with the eigenmodes of a finite line of defects in an infinite square lattice are examined.

We shall develop the analysis as follows. In Section 4.3.1, the problem of a finite line of defects (created by a perturbation of point masses) embedded in an infinite square lattice is considered. Several representations for the Green's matrix are presented, including integral forms and representation in terms of generalised hypergeometric functions. Localised defect modes for the finite line are analysed in Section 4.3.1.1. Therein, the necessary and sufficient condition for the existence of localised modes is formulated, and asymptotic expansions in the far field are also presented. Band edge expansions are constructed using an analytic continuation of the Green's function. Illustrative examples for a finite number of defects are given in Section 4.3.2, where eigenfrequencies and eigenmodes are presented and compared with the asymptotic results from the previous section. Here, the defects are characterised by one or more lattice nodes having a mass smaller than the nodes in the ambient lattice. For one- and three-dimensional multi-atomic lattices, there exists some lower bound on the difference in mass between the defect and ambient nodes such that a localised mode may be initiated [105]. However, in the present chapter, it is demonstrated that this is not the case for two-dimensional lattices: there is no lower bound on the mass that should be removed from a defect node in order to initiate a localised mode. The analysis of a finite-sized defect region is accompanied by the waveguide modes that may exist in a lattice containing an infinite chain of point masses. A brief discussion of the infinite waveguide problem is presented, for completeness, in Section 4.3.3. Finally, in Section 4.3.4, a numerical simulation illustrates that the solution for the problem of the infinite chain can be used to predict the range of eigenfrequencies of localised modes for a finite but sufficiently long array of masses representing a rectilinear defect in a square lattice.

4.3.1 Finite inclusion in an infinite square lattice

Consider a square meshing of $\mathbb{R}^2$ such that each node is labelled by the double index $\bm{n} \in \mathbb{Z}^2$, where $\bm{n} = (n_1, n_2)$. Let there be $N > 0$ defects (with $N \in \mathbb{N}$) distributed along $n_2 = 0$ as shown in Figure 4.13. The defects are characterised by a non-dimensional mass $0 < r < 1$, where the mass of the ambient nodes is taken as a natural unit. The stiffness and lengths of the lattice bonds are uniform and taken as further natural units. All physical quantities, such as the frequency and displacement, have been normalised according to these natural units and are therefore dimensionless. Let $u_{\bm{n}}$ denote the complex amplitude of the time-harmonic out-of-plane displacement of node $\bm{n}$. Then, the equation of motion is

$$u_{\bm{n}+\bm{e}_1} + u_{\bm{n}-\bm{e}_1} + u_{\bm{n}+\bm{e}_2} + u_{\bm{n}-\bm{e}_2} + (\omega^2 - 4)u_{\bm{n}}$$
$$= (1-r)\omega^2 \delta_{0,n_2} \sum_{p=0}^{N-1} u_{\bm{n}} \delta_{p,n_1}, \quad (4.39)$$

where ω is the radian frequency, $\bm{e}_i = (\delta_{1i}, \delta_{2i})^{\mathrm{T}}$, and δ_{ij} is the Kronecker delta. By means of the discrete Fourier transform the governing equation (4.39) can be written

$$(\omega^2 - 4 + 2\cos\xi_1 + 2\cos\xi_2)u^{\mathrm{FF}}(\bm{\xi}) = (1-r)\omega^2 \sum_{p=0}^{N-1} u_{p,0} \exp(-ip\xi_1). \quad (4.40)$$

Dynamic response of elastic lattices and discretised elastic membranes 117

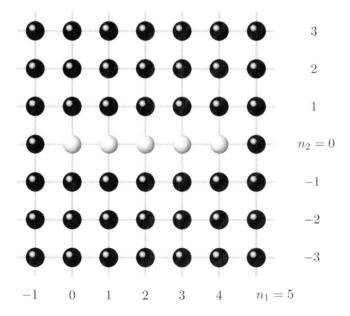

FIGURE 4.13: A finite line of defects in an infinite square lattice. The length of the links, the stiffness of the bonds and the mass of the black nodes are taken as natural units.

In the notation of previous sections, the parenthesised term on the left-hand side is $\sigma(\pmb{\xi};\omega)$ and the right-hand side of Equation (4.40) is the Fourier transform of the load $f_{N-1}(\xi_1)$. The positive root of the parenthesised term represents the dispersion equation for the ambient lattice. As mentioned earlier in Section 4.2.1.1, it is observed that for $\omega^2 > 8$ there exist no real solutions to the dispersion equation. Hence, the ambient lattice possesses a semi-infinite stop band: $\omega^2 \in (8,\infty)$. Inverting the transform yields the discrete field

$$u_{\pmb{n}}(\omega) = (1-r)\omega^2 \sum_{p=0}^{N-1} u_{p,0}\, g(\pmb{n},p;\omega), \qquad (4.41)$$

where $g(\pmb{n},p;\omega)$ is the shifted lattice Green's function defined as:

$$g(\pmb{n},p;\omega) = \frac{1}{\pi^2} \int_0^\pi \int_0^\pi \frac{\cos([n_1-p]\xi_1)\cos(n_2\xi_2)}{\omega^2 - 4 + 2\cos\xi_1 + 2\cos\xi_2}\, d\xi_1\, d\xi_2. \qquad (4.42)$$

For the purposes of numerical evaluation and asymptotic analysis in the stop band of the ambient lattice ($\omega^2 > 8$), it is convenient to rewrite the lattice Green's function as a single integral

$$g(\pmb{n},p;\omega) = \frac{1}{2\pi} \int_0^\pi \frac{(\sqrt{a^2-1}-a)^{|n_1-p|}}{\sqrt{a^2-1}} \cos(n_2\xi_2)\, d\xi_2, \qquad (4.43)$$

where $a(\xi_2;\omega) = \omega^2/2 - 2 + \cos\xi_2$. Reversing the order of integration yields the same result, but with $n_1 - p$ and n_2 interchanged, and ξ_1 interchanged with ξ_2. An alternative representation can be found in [184]:

$$g(\boldsymbol{n},p;\omega) = \frac{(-1)^{n_1-p+n_2}}{2} \int_0^\infty I_{n_1-p}(x)I_{n_2}(x)e^{-\alpha x}\,dx, \qquad (4.44)$$

where $I_m(x)$ is the modified Bessel function of the first kind, $\alpha = \omega^2/2-2 > 2$. The integral is symmetric about $n_1 - p = 0$ and $n_2 = 0$ and therefore it may be assumed, without loss of generality, that $n_1 \geq p$ and $n_2 \geq 0$. The integral (4.44) may then be represented in terms of regularised generalised hypergeometric functions (see [167], Section 3.15.6, Equation 8)

$$g(\boldsymbol{n},p;\omega) = \frac{(-1)^{m+n_2}}{(2\alpha)^{1+m+n_2}}((m+n_2)!)^2 \; {}_4\mathbf{F}_3\left[\begin{matrix} a_1,\, a_1,\, a_2,\, a_2 \\ b_1,\, b_2,\, b_1+b_2-1 \end{matrix}; \frac{4}{\alpha^2}\right], \qquad (4.45)$$

where $m = n_1 - p$, $a_1 = (1+m+n_2)/2$, $a_2 = (2+m+n_2)/2$, $b_1 = 1+m$, and $b_2 = 1 + n_2$. The series (4.45) is convergent for $\alpha^2 > 4$ (see [148]), that is, everywhere in the stop band of the ambient lattice. It is observed that along the ray $m = n_2$, the Green's function may be written in terms of Gauss' hypergeometric function. In particular, Equation (4.45) reduces to

$$g(n,n,0;\omega) = \frac{((2n)!)^2}{(2\alpha)^{1+2n}} \; {}_2\mathbf{F}_1\left[\begin{matrix} 1/2+n,\, 1/2+n \\ 1+2n \end{matrix}; \frac{4}{\alpha^2}\right]. \qquad (4.46)$$

The function (4.46) is strictly positive in the region $n \geq 0$ and $\alpha > 2$. Hence, for a single defect, the lattice nodes along the diagonal rays do not oscillate relative to each other.

Furthermore, for the case of $m = n_2 = 0$, the integral representation (4.42) reduces to the twofold Watson integral (see, for example, [84] and [191]). Using a simple change of variables (4.42) can be written in terms of an elliptic integral, or alternatively, one can use (4.46) and observe that

$$g(0,0;\omega) = \frac{1}{2\alpha} \; {}_2\mathbf{F}_1\left[\begin{matrix} 1/2,\, 1/2 \\ 1 \end{matrix}; \frac{4}{\alpha^2}\right] = \frac{1}{\alpha\pi}K\left(\frac{4}{\alpha^2}\right), \qquad (4.47)$$

where $K(x)$ is the complete elliptical integral of the first kind. Together with Equation (4.47), the representation (4.44) is particularly useful since, by repeated integration by parts and use of the identity $I_n(x) = 2I'_{n-1}(x) - I_{n-2}(x)$, one can iterate from $g(0,0;\omega)$ to a general $g(\boldsymbol{n},p;\omega)$.

4.3.1.1 Localised modes

Of primary interest are localised modes, that is, modes of vibration at frequencies that are not supported by the ambient lattice and therefore decay rapidly away from the defect. Introducing the vector $\mathcal{U} = $

$(u_{0,0}, u_{2,0}, \ldots, u_{N-1,0})^{\mathrm{T}}$ and choosing $n_2 = 0$ in Equation (4.41) yields the eigenvalue problem
$$\mathcal{U} = (1-r)\omega^2 \mathcal{G}(\omega)\mathcal{U}, \tag{4.48}$$
where the matrix entries $(\mathcal{G}(\omega))_{ij} = g(i-1, 0, j-1; \omega)$. Clearly, $\mathcal{G}$ is symmetric and Toeplitz (and hence bisymmetric and centrosymmetric)
$$\mathcal{G} = \begin{pmatrix} \mathcal{G}_{11} & \mathcal{G}_{12} & \mathcal{G}_{13} & \cdots & \mathcal{G}_{1(N-1)} & \mathcal{G}_{1N} \\ & \mathcal{G}_{11} & \mathcal{G}_{12} & \cdots & \mathcal{G}_{1(N-2)} & \mathcal{G}_{1(N-1)} \\ & & \mathcal{G}_{11} & \cdots & \mathcal{G}_{1(N-3)} & \mathcal{G}_{1(N-2)} \\ & & & \ddots & \vdots & \vdots \\ & & & & \mathcal{G}_{11} & \mathcal{G}_{12} \\ & & & & & \mathcal{G}_{11} \end{pmatrix}, \tag{4.49}$$
which greatly reduces the number of required computations. Indeed, for N defects the matrix $\mathcal{G}$ has N independent elements. The solvability condition of the spectral problem (4.48) yields a transcendental equation in ω,
$$\det\left(\mathbb{I}_N - (1-r)\omega^2 \mathcal{G}\right) = 0, \tag{4.50}$$
where $\mathbb{I}_N$ is the $N \times N$ identity matrix. Equation (4.50) is the necessary and sufficient condition for the existence of localised modes. Symmetry implies that there exists an orthonormal set of N eigenvectors of $\mathcal{G}$ and hence, N eigenvalues (frequencies). The centrosymmetry of $\mathcal{G}$ allows the number of symmetric and skew-symmetric modes to be determined (see, for example, [38]). Introducing the $N \times N$ exchange matrix
$$\mathbb{J}_N = \begin{pmatrix} 0 & 0 & 0 & 1 \\ 0 & 0 & 1 & 0 \\ 0 & \cdot^{\cdot^{\cdot}} & 0 & 0 \\ 1 & 0 & 0 & 0 \end{pmatrix}, \tag{4.51}$$
an eigenmode is said to be symmetric if $\mathcal{U} = \mathbb{J}_N \mathcal{U}$ and skew-symmetric if $\mathcal{U} = -\mathbb{J}_N \mathcal{U}$. For a system of N defects there exist $\lceil N/2 \rceil$ symmetric modes and $\lfloor N/2 \rfloor$ skew-symmetric modes, where $\lceil \cdot \rceil$ and $\lfloor \cdot \rfloor$ are the ceiling and floor operators respectively. Of course here, symmetry refers to the symmetry of the eigenmodes in the n_1-direction about the centre of the defect line. Due to the symmetry of the system, all modes are symmetric about the line $n_2 = 0$.

Consider the total force on an inclusion containing N defects
$$F = \sum_{p=0}^{N-1} (u_{p-1,0} + u_{p+1,0} + 2u_{p,1} - 4u_{p,0}). \tag{4.52}$$
By definition, for a skew-symmetric mode $u_{p,0} = -u_{N-1-p,0}$ and further $u_{p,q} = -u_{N-1-p,q}$. Hence, for all skew-symmetric modes the inclusion is self-balanced (i.e. $F = 0$) and therefore, all skew-symmetric localised modes can be considered as multipole modes.

For the illustrative examples presented later, the eigenvalue problem (4.48) will be solved for the unit eigenvectors ($\|\mathcal{U}\| = 1$).

4.3.1.2 Asymptotic expansions in the far field

Here, asymptotic expansions in the far field are considered for some particular cases. Asymptotic expansions for an isolated Green's matrix in various configurations have already been considered in Section 4.1.1 and the approach detailed previously is used again here to construct expansions of the shifted lattice Green's function.

Far field, along the line of defects. The case of $n_1 \to \infty$, $n_2 = 0$ and finite N is considered. Introducing the small parameter $\varepsilon = p/n_1$, the kernel of (4.43) may be expanded for small $|\varepsilon| \ll 1$. In particular,

$$\left(\sqrt{a^2-1}-a\right)^{|n_1-p|} \sim \left(\sqrt{a^2-1}-a\right)^{|n_1|}\left[1-\varepsilon\log\left(\sqrt{a^2-1}-a\right)\right]^{|n_1|}. \tag{4.53}$$

It is observed that at large n_1 and sufficiently small N, the dominant contribution to the integral (4.43) comes from a small region in the vicinity of $\xi_2 = \pi$. Therefore,

$$\left(\sqrt{a^2-1}-a\right)^{|n_1-p|} \sim \left(\sqrt{c^2-1}-c\right)^{|n_1|}\left[1-\frac{(\pi-\xi_2)^2}{2\sqrt{c^2-1}}\right]^{|n_1|}$$
$$\times \left[1-\varepsilon\log\left(\sqrt{c^2-1}-c\right)+\varepsilon\frac{(\pi-\xi_2)^2}{2\sqrt{c^2-1}}\right]^{|n_1|}, \tag{4.54}$$

where $c = \omega^2/2 - 3$. Thus,

$$\left(\sqrt{a^2-1}-a\right)^{|n_1-p|} \sim \left(\sqrt{c^2-1}-c\right)^{|n_1-p|}\exp\left[-|n_1-p|\frac{(\pi-\xi_2)^2}{2\sqrt{c^2-1}}\right]. \tag{4.55}$$

In addition, $1/\sqrt{a^2-1} \sim 1/\sqrt{c^2-1}$. Hence, for $0 < \varepsilon \ll 1$ and making use of (4.43)

$$g(n_1,0,p;\omega) \sim \frac{\left(\sqrt{c^2-1}-c\right)^{|n_1-p|}}{2\pi\sqrt{c^2-1}} \int_{\pi-\varepsilon}^{\pi} \exp\left[-|n_1-p|\frac{(\pi-\xi_2)^2}{2\sqrt{c^2-1}}\right] d\xi_2. \tag{4.56}$$

Making the substitution $x = (\pi-\xi_2)\sqrt{|n_1-p|/(2\sqrt{c^2-1})}$, and performing the resulting integration yields

$$g(n_1,0,p;\omega) \sim \frac{\left(\sqrt{c^2-1}-c\right)^{|n_1-p|}}{\sqrt{8\pi\sqrt{c^2-1}}} \frac{1}{\sqrt{|n_1-p|}} \quad \text{as} \quad n_1 \to \infty. \tag{4.57}$$

Thus, from (4.41), the physical field has the following approximate representation for $n_1 \to \infty$

$$u_{n_1,0}(\omega) \sim (1-r)\omega^2 \sum_{p=0}^{N-1} \frac{\left(\sqrt{c^2-1}-c\right)^{|n_1-p|}}{\sqrt{8\pi\sqrt{c^2-1}}} \frac{u_{p,0}(\omega)}{\sqrt{|n_1-p|}}, \tag{4.58}$$

where $u_{p,0}(\omega)$ should be determined from (4.48). It is observed that, when $N = 1$, Equation (4.58) is consistent with Equation (4.17) of [131] up to a change in sign.

Far field, perpendicular to the line of defects. Here, the case considered is $n_1 = p'$, $n_2 \to \infty$ with N and p' finite. The kernel is oscillatory and is approximated as a product of decaying and oscillatory functions. For sufficiently small $|p'-p|$ and large n_2, the non-oscillatory part of the integrand in (4.43) is approximated as before, leading to

$$g(p', n_2, p; \omega) \sim \frac{\left(\sqrt{c^2-1}-c\right)^{|n_2|}}{2\pi\sqrt{c^2-1}}$$
$$\times \int_{\pi-\varepsilon}^{\pi} \exp\left[-|n_2|\frac{(\pi-\xi_1)^2}{2\sqrt{c^2-1}}\right]\cos\left([p'-p]\xi_1\right)d\xi_1. \quad (4.59)$$

Making a similar change of variable, $x = (\pi - \xi_1)\sqrt{|n_2|/(2\sqrt{c^2-1})}$, and integrating, it is found that

$$g(p', n_2, p; \omega) \sim (-1)^{(p'-p)}\frac{\left(\sqrt{c^2-1}-c\right)^{|n_2|}}{\sqrt{8\pi\sqrt{c^2-1}}}\frac{1}{\sqrt{|n_2|}}$$
$$\times \exp\left[-(p'-p)^2\frac{\sqrt{c^2-1}}{2|n_2|}\right]. \quad (4.60)$$

Hence, for $n_2 \to \infty$ the physical field in (4.41) may be approximated by

$$u_{p', n_2}(\omega) \sim (1-r)\omega^2 \frac{\left(\sqrt{c^2-1}-c\right)^{|n_2|}}{\sqrt{8\pi\sqrt{c^2-1}}}$$
$$\times \sum_{p=0}^{N-1}(-1)^{(p'-p)}\exp\left[-(p'-p)^2\frac{\sqrt{c^2-1}}{2|n_2|}\right]\frac{u_{p,0}(\omega)}{\sqrt{|n_2|}}. \quad (4.61)$$

It is observed that for $N = 1$ and $p' = p$, the above Equation (4.61) is consistent with Equation (4.17) of [131] up to a change in sign. Moreover, for the case of $p' = p$, (4.61) reduces to (4.58).

4.3.1.3 Band edge expansions

The representations of Green's matrix (4.43)–(4.45) are valid in the stop band. However, given that the hypergeometric function in the representation (4.45) is zero balanced, that is, the sum of the bottom parameters minus the sum of the top parameters vanishes: $2(b_1 + b_2) - 1 - 2(a_1 + a_2) = 0$, the stop band Green's matrix can be extended to the boundary of the pass band by analytic continuation.* In particular, the analytical continuation of

---
*Indeed, for any integer balanced hypergeometric function $_{q+1}F_q$ there exists an analytic continuation to the boundary of the unit disk (see [33], among others, for details).

the function (4.45) has the form

$$g(\boldsymbol{n},p;\omega) = \frac{(-4)^{m+n_2}}{\pi(2\alpha)^{1+m+n_2}} \sum_{j=0}^{\infty} \left(\frac{([1+m+n_2]/2)_j}{j!}\right)^2 \left(1-\frac{4}{\alpha^2}\right)^j$$

$$\times \left\{ \sum_{k=0}^{j} \frac{(-j)_k}{\{([1+m+n_2]/2)_j\}^2} \mathcal{H}(m,n_2,k) \left[\psi(1+j-k) \right. \right.$$

$$+\psi(1+j) - \psi\left(\frac{1+m+n_2}{2}+j\right) - \log\left(1-\frac{4}{\alpha^2}\right)\right]$$

$$\left. +(-1)^j (j)! \sum_{k=j+1}^{\infty} \frac{(k-j-1)!}{\{([1+m+n_2]/2)_k\}^2} \mathcal{H}(m,n_2,k) \right\}, \quad (4.62)$$

where the reader is reminded that $m = n_1 - p$, $(\cdot)_j$ is the Pochhammer symbol, $\psi(x)$ is the Digamma function, and

$$\mathcal{H}(m,n,k) = \frac{(m)_k (n)_k}{k!} \, {}_3F_2\left[\begin{matrix}(m+n_2)/2,\ (m+n_2)/2,\ -k\\ m,\ n\end{matrix};1\right]. \quad (4.63)$$

The symbol ${}_pF_q[\ldots]$ denotes the generalised hypergeometric function (see, for example, [33]), which is related to the regularised generalised hypergeometric function thus:

$${}_pF_q[a_1,\ldots,a_p;b_1,\ldots b_q;z] = \{\Gamma(b_1)\ldots\Gamma(b_q)\}\,{}_p\mathbf{F}_q[a_1,\ldots,a_p;b_1,\ldots b_q;z].$$

In this case, the continuation (4.62) holds for $\alpha^2 \geq 4$, which in terms of frequency corresponds to $\omega^2 \geq 8$. It is emphasised that in this section, the term "*vicinity of the band edge*" refers to a small interval $8 \leq \omega^2 < 8+\varepsilon$, where $0 < \varepsilon \ll 1$.

Hence, choosing $j = 0$ yields the leading order behaviour of (4.45) as $\alpha^2 \to 4^+$ ($\omega^2 \to 8^+$), that is, as ω approaches the boundary of the pass band from the stop band:

$$g(\boldsymbol{n},p;\omega) \sim \frac{(-4)^{m+n_2}}{\pi(2\alpha)^{1+m+n_2}} \left\{ \left[-2\gamma - \psi\left(\frac{1+m+n_2}{2}\right) - \log\left(1-\frac{4}{\alpha^2}\right)\right] \right.$$

$$\left. + \sum_{k=1}^{\infty} \frac{(k-1)!}{\{([1+m+n_2]/2)_k\}^2} \mathcal{H}(m,n_2,k) \right\}, \quad (4.64)$$

where γ is the Euler–Mascheroni constant. Alternative representations of the leading order continuations for general zero-balanced ${}_{q+1}F_q$ were derived by Saigo and Srivastava [171]. Since $k > 0$, the series representation of the hypergeometric function in (4.63) has a finite number of terms and therefore may be computed exactly. The convergence condition for the infinite sums in (4.62)

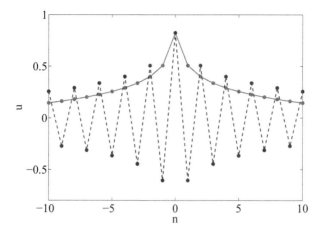

FIGURE 4.14: The solid curve shows the asymptotic expression for the displacement field along the diagonal $n_1 = n_2$ (with $p = 0$) in the vicinity of the band edge (see Equation (4.65b)). The dashed curve shows the corresponding asymptotic expression for the field along the bond line (see Equation (4.65a)). The frequency chosen is $\omega = 2.829$.

and (4.64) is $2 + m + n_2 + j > 0$, and is automatically satisfied since it was assumed (without loss of generality) at the outset that $m \geq 0$ and $n_2 \geq 0$.

The asymptotic expression (4.64) is particularly interesting as it elucidates the nature of the singularity of the lattice Green's matrix at the band edge. In particular, the asymptotic representation (4.64) captures the logarithmic singularity as $\omega^2 \to 8^+$. This logarithmically singular behaviour near the band edge is not obvious from the original representations presented earlier (see equations (4.43)–(4.45)).

For some particular cases, Equation (4.64) reduces to the following simplified forms. Along the rays[†] $m = 0$ (i.e. $n_1 = p$):

$$g(p, n_2, p; \omega) \sim \frac{(-4)^{1+n_2}}{\pi(2\alpha)^{1+n_2}} \left[2\gamma + \psi\left(\frac{1+n_2}{2}\right) \right.$$
$$\left. + \log\left(1 - \frac{4}{\alpha^2}\right) \right] \triangleq \tilde{g}^{(\text{bond})}(n_2; \omega), \quad (4.65a)$$

[†]Equivalently, one may substitute n_2 by $n_1 - p$ in (4.65a) to obtain expansions of g along $n_2 = 0$.

and along the diagonal rays $m = n_2$:

$$g(n_1, m, p; \omega) \sim -\frac{16^m}{\pi(2\alpha)^{1+2m}}\left[2\gamma + \psi\left(\frac{1}{2}+m\right)\right.$$
$$\left.+ \log\left(1 - \frac{4}{\alpha^2}\right)\right] \triangleq \tilde{g}^{(\text{diag})}(m; \omega). \quad (4.65b)$$

The Digamma function grows logarithmically as $m \to \infty$ and the term $2\gamma + \psi(1/2+m)$ is strictly positive for $m > 0$. Therefore, for sufficiently small m the bracketed term in Equations (4.65) is negative in the neighbourhood of $\alpha = 2$. Hence, in the vicinity of the band edge, the stop band Green's matrix exhibits fundamentally different behaviour along the bond lines compared with the diagonal rays. In particular, along the bond lines the masses will oscillate out of phase, whereas for the diagonal ray lines the masses will oscillate in phase, as illustrated in Figure 4.14. In the far field, Equations (4.65) further reduce to

$$g(p, n_2, p; \omega) \sim \frac{(-4)^{1+n_2}}{\pi(2\alpha)^{1+n_2}}\left[2\gamma + \log\left(\frac{n_2}{2}\right) + \log\left(1 - \frac{4}{\alpha^2}\right)\right], \text{ as } n_2 \to \infty,$$
$$(4.66a)$$

$$g(m, m, p; \omega) \sim -\frac{16^m}{\pi(2\alpha)^{1+2m}}\left[2\gamma + \log m + \log\left(1 - \frac{4}{\alpha^2}\right)\right], \text{ as } m \to \infty.$$
$$(4.66b)$$

Using Equations (4.41) and (4.65) the out-of-plane displacement for a lattice with N defects has the following asymptotic representation in the vicinity of the band edge

$$u_{n_1,0}(\omega) \sim (1-r)\omega^2 \sum_{p=0}^{N-1} u_{p,0}\, \tilde{g}^{(\text{bond})}(n_1 - p; \omega), \text{ as } \omega^2 \to 8^+, \quad (4.67a)$$

$$u_{n_1,n_2-p}(\omega) \sim (1-r)\omega^2 \sum_{p=0}^{N-1} u_{p,0}\, \tilde{g}^{(\text{diag})}(n_1 - p; \omega), \text{ as } \omega^2 \to 8^+, \quad (4.67b)$$

along the rays $n_2 = 0$ and $n_2 = n_1 - p$, respectively.

4.3.2 Illustrative examples

Several particular cases are considered here corresponding to relatively short defects with $N \in \{1, 2, 3\}$. The solid curves in Figure 4.15 show the ith solution, $r_{N,i}(\omega)$, of the solvability condition (4.50) for a line of N defects. The shaded region indicates the stop band ($\omega^2 > 8$) of the ambient lattice. For frequencies in this region, waves in the ambient lattice will decay exponentially away from the defect or source.

It is interesting to note that, for one- and three-dimensional multi-atomic lattices, there exists some lower bound on the amount of mass that should be removed from the defect nodes such that a localised mode may be initiated (see, for instance, [105]). However, here the image of $r_{N,N}(\omega)$, indicated by the solid curves in Figure 4.15, is $(0, 1)$. In other words, there is no lower bound on the mass that should be removed from a defect node in order to initiate a localised mode. As $r \to 1$, that is, as the lattice approaches a homogeneous lattice, the frequency of the localised mode approaches the band edge ($\omega^2 \to 8^+$). It is also observed that for $N > 1$, the solid curves intersect the band edge at several distinct values of r. This suggests that for a given number of defects, there exists a maximum value of r below which all possible localised eigenmodes may be initiated. Above this value of r it is only possible to initiate a subset of the possible eigenmodes with the lower frequency eigenmodes being filtered out. In all cases, the highest frequency eigenmode persists for all possible values of r on $(0, 1)$. For fixed ω, the solvability condition (4.50) for a system of N defects is a polynomial in r of at most degree N. Therefore, there exist no more than N solutions for a given frequency ω.

The dashed curves correspond to the problem of an isolated chain of N particles of non-dimensional mass r^*, connected by springs to two nearest neighbours and surrounded by rigid foundations. For such a problem, the out-of-plane displacement of mass $n \in \mathbb{Z}$ satisfies

$$\mathcal{L}(v_0, v_1, \cdots, v_{N-1})^{\mathrm{T}} = 0, \tag{4.68}$$

where the matrix $\mathcal{L}$ has elements

$$(\mathcal{L})_{ij} = (r^*\omega^2 - 4)\delta_{ij} + \delta_{i-1,j} + \delta_{i,j-1}. \tag{4.69}$$

The dashed curves in Figure 4.15 represent the solutions $r^*_{N,i}(\omega)$ of the solvability condition: $\det \mathcal{L} = 0$. It is observed that as $\omega \to \infty$, the dashed curves approach the solid curves from below.

4.3.2.1 Single defect

For the case of a single defect located at the origin, the quantity $\mathcal{G}$ in (4.48) is a scalar:

$$\mathcal{G}(\omega) = \frac{1}{\alpha\pi} K\left(\frac{4}{\alpha^2}\right), \tag{4.70}$$

where $K(x)$ is the complete elliptical integral of the first kind. The solvability condition may be written as

$$r_{1,1} = 1 + \pi\left(\frac{2}{\omega^2} - \frac{1}{2}\right)\left[K\left(\frac{16}{(\omega^2 - 4)^2}\right)\right]^{-1}, \tag{4.71}$$

which has the leading order asymptotic representation

$$r_{1,1} \sim \frac{4}{\omega^2}, \quad \text{as} \quad \omega \to \infty. \tag{4.72}$$

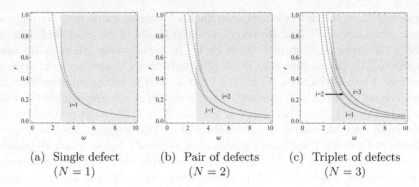

(a) Single defect ($N = 1$) (b) Pair of defects ($N = 2$) (c) Triplet of defects ($N = 3$)

FIGURE 4.15: The solid curves show the ith solution, $r_{N,i}(\omega)$, of the solvability condition (4.50) for a system of N defects embedded in the square lattice. The shaded region ($\omega^2 > 8$) indicates the stop band of the ambient lattice. The dashed curves show the corresponding i^{th} solution, $r^*_{N,i}(\omega)$, of the solvability condition for an isolated system of N defects (see Equation (4.68)).

It is observed that the solvability condition for Equation (4.68) with $N = 1$ agrees precisely with the leading-order high-frequency asymptotic expansion, hence, the observed coalescence of the solid and dashed curves in Figure 4.15a.

The localised defect mode is shown in Figure 4.16a, together with field along the line $n_2 = 0$ and the associated asymptotic field as $n_1 \to \infty$ in Figure 4.16b. Figures 4.16c and 4.16d show the field (solid line) and the band edge asymptotics (dashed line) for a value of $\alpha = 2.006$. The asymptotic expansions show good agreement with the computed field, even for the far field asymptotics in the neighbourhood of the defect.

4.3.2.2 Pair of defects

In the case of a pair of defects, $\mathcal{G}(\omega)$ is a 2×2 matrix with the diagonal elements given by (4.70). The off-diagonal elements have the form

$$(\mathcal{G}(w))_{12} = (\mathcal{G}(w))_{21} = \frac{1}{4} - \frac{1}{2\pi} K\left(\frac{4}{\alpha^2}\right). \quad (4.73)$$

The solutions of the solvability condition are

$$r_{2,1} = 1 - \frac{4\pi(\omega^2 - 4)}{\pi\omega^2(\omega^2 - 4) - 2\omega^2(\omega^2 - 8)K\left(\frac{16}{(\omega^2 - 4)^2}\right)}, \quad (4.74)$$

$$r_{2,2} = 1 + \frac{4\pi(\omega^2 - 4)}{\pi\omega^2(\omega^2 - 4) - 2\omega^4 K\left(\frac{16}{(\omega^2 - 4)^2}\right)}, \quad (4.75)$$

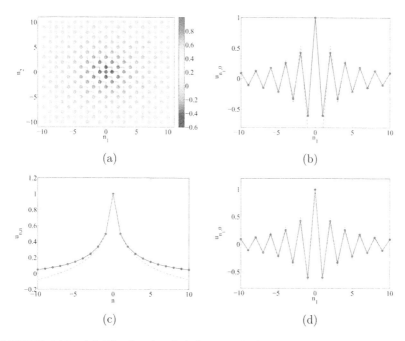

FIGURE 4.16: (a) The localised defect mode for a single defect with $r = 0.8$ and $\omega = 2.83$. (b) The solid curve is the out-of-plane displacement along the line $n_2 = 0$ and the dashed curve is the asymptotic expansion for $n_1 \to \infty$ (see Equation (4.58)). (c) The out-of-plane displacement along the line $n_1 = n_2$ (solid curve) with the corresponding asymptotic expansion (4.65b) for the band edge (dashed curve). (d) As for (b), but the dashed curve represents the band edge expansion along $n_2 = 0$ (see Equation (4.67a)).

whence the leading-order high-frequency asymptotic expansions are

$$r_{2,1} \sim \frac{3}{\omega^2} \quad \text{and} \quad r_{2,2} \sim \frac{5}{\omega^2} \quad \text{as} \quad \omega \to \infty, \qquad (4.76)$$

which agree with the solvability condition of the isolated system (4.68) for $N = 2$, hence, the observed coalescence of the solid and dashed curves in Figure 4.15.

Figure 4.17 shows the two defect modes together with the field along the lines $n_1 = 0$, and $n_2 = 0$ and the associated asymptotic field at infinity. In addition, the dash-dot line in Figure 4.17c shows the band edge expansion in the vicinity of $\alpha = 2$. In this case, Figure 4.17c corresponds to value of $\alpha \approx 2.025$. Once again, the asymptotics are in good agreement with the computed field. Due to the symmetry, the field along the line $n_1 = 1$ is identical to that in Figure 4.17e for the symmetric case and identical up to a reflection in the line $u_{0,n_2} = 0$ in Figure 4.17f for the skew-symmetric case.

The lower solid curve in Figure 4.15b corresponds to $r_{2,1}$ as defined

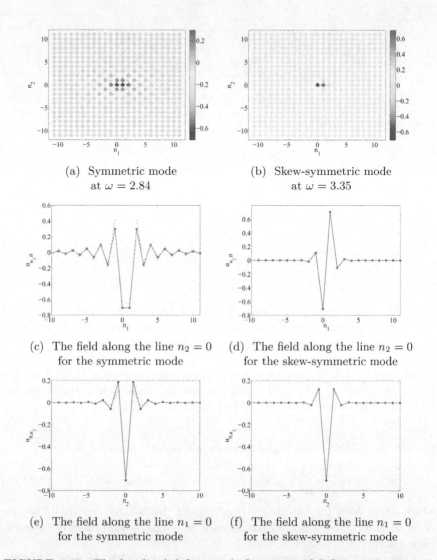

FIGURE 4.17: The localised defect mode for a pair of defects with $r = 0.49$. The solid curves show the out-of-plane displacement along the indicated line, and the dashed curves are the associated asymptotic expansions in the far field (see Equations (4.58) and (4.61) as appropriate). The dash-dot curve in Figure 4.17c shows the band edge expansion (see Equation (4.67a)).

in (4.74). The maximum value of the lower solid curve is given by

$$r_{2,1}^{(\max)} = \lim_{\omega^2 \to 8^+} r_{2,1} = \frac{1}{2}. \quad (4.77)$$

Hence for a pair of defects, a symmetric localised mode cannot be initiated for $r \geq 1/2$.

4.3.2.3 Triplet of defects

For the case of three defects, the 3×3 matrix $\mathcal{G}(\omega)$ has the $(\mathcal{G})_{11}$ and $(\mathcal{G})_{12}$ elements as defined in Equations (4.70) and (4.73). The remaining independent component is

$$(\mathcal{G}(w))_{13} = (\mathcal{G}(\omega))_{11} - \frac{\alpha}{2} + \frac{\alpha}{\pi} E\left(\frac{4}{\alpha^2}\right), \quad (4.78)$$

where $E(x)$ is the complete Elliptic Integral of the second kind. The solutions of the solvability condition are of similar form to the previous two cases and are omitted for brevity. The high frequency asymptotics for $r(\omega)$ are

$$r_{3,1} \sim \frac{4 - \sqrt{2}}{\omega^2}, \quad r_{3,2} \sim \frac{4}{\omega^2}, \quad \text{and} \quad r_{3,3} \sim \frac{4 + \sqrt{2}}{\omega^2} \quad \text{as} \quad \omega \to \infty, \quad (4.79)$$

which again coincide with the solvability condition for (4.68) for the case of a particle triplet ($N = 3$). The maximum values of $r_{3,i}(\omega)$ are $r_{3,1}^{(\max)} = 1 - 3\pi/16$, $r_{3,2}^{(\max)} = 7/8 - (8 - 4\pi)^{-1}$, and $r_{3,3}^{(\max)} = 1$.

The three localised eigenmodes, along with plots of the associated asymptotic expressions are shown in Figures 4.18–4.20 for a contrast ratio of $r = 0.4$. Plots of the displacement field along the lines $n_2 = 0$, $n_1 = 1$ and $n_1 = 0$ are also provided together with their associated asymptotic fields. In each case, the solid curves show the displacement field, whilst the dashed curves show the associated asymptotics in the far field. The dash-dot line in Figure 4.18b shows the band edge expansion in the vicinity of $\alpha = 2$. In this case, Figure 4.18b corresponds to value of $\alpha \approx 2.017$. There are two symmetric modes (the lowest and highest frequency modes) and a single skew-symmetric mode, as expected from the properties of $\mathcal{G}$ discussed in the previous subsection. However, for defects of mass $r \geq r_{3,1}^{(\max)}$, it is not possible to initiate the lower frequency symmetric eigenmode and only a further symmetric mode and a skew-symmetric mode persist. For values of $r \geq r_{3,2}^{(\max)}$, it is only possible to initiate the highest frequency symmetric mode.

4.3.3 Infinite line of defects in a square lattice

The section will be devoted to the discussion of an infinite line of defects embedded in an infinite square lattice, as shown in Figure 4.21. As in the previous section, the defects are characterised by a non-dimensional mass $0 < r < 1$. A recent paper by Osharovich and Ayzenberg-Stepanenko [152]

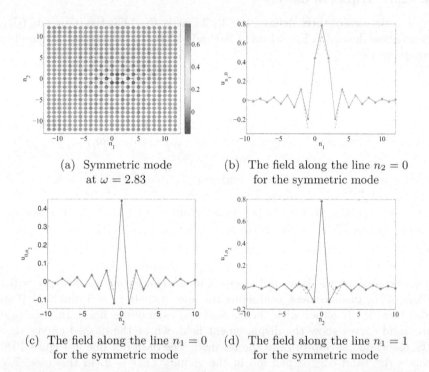

(a) Symmetric mode at $\omega = 2.83$

(b) The field along the line $n_2 = 0$ for the symmetric mode

(c) The field along the line $n_1 = 0$ for the symmetric mode

(d) The field along the line $n_1 = 1$ for the symmetric mode

FIGURE 4.18: The first localised defect mode for a triplet of defects with $r = 0.4$. The solid curves show the out-of-plane displacement along the indicated line, and the dashed curves are the associated asymptotic expansions in the far field (see Equations (4.58) and (4.61) as appropriate). The dash-dot line in (b) corresponds to the band edge expansion (see Equation (4.67a)).

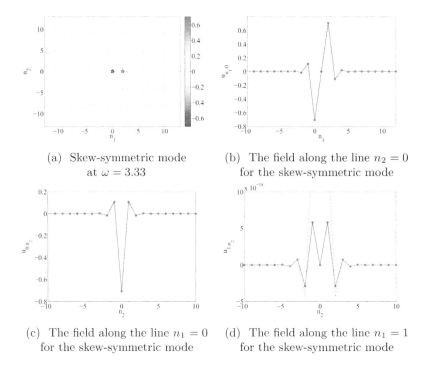

FIGURE 4.19: The second localised mode for a triplet of defects. The solid curves show the out-of-plane displacement along the indicated line, and the dashed curves are the associated asymptotic expansions in the far field (see Equations (4.58) and (4.61) as appropriate).

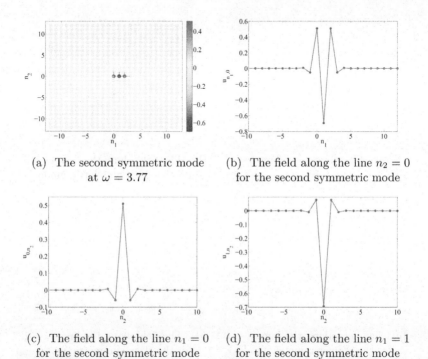

FIGURE 4.20: The third localised defect mode for a triplet of defects. The solid curves are the out-of-plane displacement along the indicated line, and the dashed curves are the associated asymptotic expansions in the far field (see Equations (4.58) and (4.61) as appropriate).

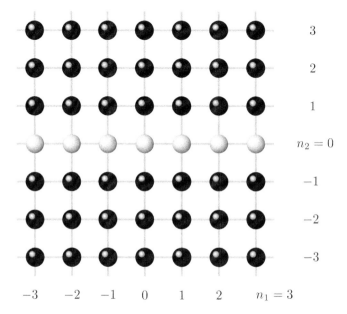

FIGURE 4.21: A square cell lattice containing an infinite chain of defects with non-dimensional mass r along $n_2 = 0$, and an ambient lattice composed of particles with unit mass. As before, the stiffness and length of the links are taken as natural units.

studied the waveguide problem for an infinite linear defect embedded in a square lattice. More recently, Colquitt et al. [53] studied in detail this precise problem.

4.3.3.1 Equations of motion

Given the symmetry of the system about the line $n_2 = 0$ (see Figure 4.21), it is convenient to reduce the problem to a half-plane system, which can be formulated as follows. The displacement amplitude field for time-harmonic disturbances in the upper-half plane, $\bm{n} \in \mathbb{Z} \times \mathbb{Z}^+$ is

$$u_{\bm{n}+\bm{e}_1} + u_{\bm{n}-\bm{e}_1} + u_{\bm{n}+\bm{e}_2} + u_{\bm{n}-\bm{e}_2} + (\omega^2 - 4)u_{\bm{n}} = 0, \tag{4.80a}$$

and for $n_1 \in \mathbb{Z}$, $n_2 = 0$ is

$$u_{n_1+1,0} + u_{n_1-1,0} + u_{n_1,1} + u_{n_1,-1} + (r\omega^2 - 4)u_{n_1,0} = 0. \tag{4.80b}$$

Taking the discrete Fourier transform in the n_1–direction yields

$$u^{\text{F}}_{n_2+1} + u^{\text{F}}_{n_2-1} - 2\Omega_1(\xi,\omega)u^{\text{F}}_{n_2} = 0, \tag{4.81a}$$

and

$$u^{\text{F}}_{1} + u^{\text{F}}_{-1} - 2\Omega_r(\xi,\omega)u^{\text{F}}_{0} = 0, \tag{4.81b}$$

where

$$\Omega_\beta(\xi,\omega) = 1 + 2\sin^2\left(\frac{\xi}{2}\right) - \frac{\beta\omega^2}{2} \tag{4.82}$$

and ξ is the Fourier parameter. For $n_2 > 1$ a solution of the form

$$u^{\text{F}}_{n_2} = \lambda^{n_2} u^{\text{F}}_1, \quad \text{with } |\lambda| \leq 1, \tag{4.83}$$

is sought. The case of $|\lambda| = 1$ corresponds to a displacement field which propagates sinusoidally, with constant amplitude, away from the defect along $n_2 = 0$. The condition $|\lambda| < 1$ corresponds to a localised mode, the amplitude of which, decays exponentially away from the waveguide along $n_2 = 0$. The primary focus of this chapter is localised modes, therefore the following discussion will be devoted to the latter case of $|\lambda| < 1$. For a detailed analysis of the system, the reader is referred to [53, §4]. Together, Equations (4.81a) and (4.83) yield an expression for the factor λ corresponding to localised modes

$$\lambda = 1 + 2\sin^2\left(\frac{\xi}{2}\right) - \frac{r\omega^2}{2}. \tag{4.84}$$

Skew-symmetric solutions. Consider solutions that are skew-symmetric about the line $n_2 = 0$. These modes satisfy the symmetry condition $u_{n_1,n_2} = -u_{n_1,-n_2}$, whence $u_{n_1,0} = 0$ and hence $u_{\bm{n}} = 0$. In other words, the only skew-symmetric solution is the trivial one.

Dynamic response of elastic lattices and discretised elastic membranes 135

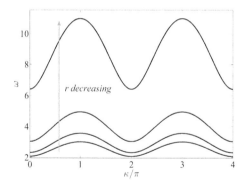

FIGURE 4.22: The quantity $\omega^{(-)}$, given in equation (4.85), plotted as a function of the normalised Bloch parameter ξ/π for $r = 0.05, 0.25, 0.5$ and 0.75.

Symmetric solutions. For the case when symmetry conditions are imposed about $n_2 = 0$, that is $u_{n_1,n_2} = u_{n_1,-n_2}$, the dispersion equation for localised defect modes supported by the infinite line defect is given by

$$\omega^{(-)}(\xi) = \left\{ \frac{2}{r(2-r)} \left[1 + 2\sin^2(\xi/2) \right. \right.$$
$$\left. \left. + \sqrt{1 + 4(1-r)^2 \sin^2(\xi/2)(1 + \sin^2(\xi/2))} \right] \right\}^{1/2}. \quad (4.85)$$

This dispersion relation is determined in two parts. First, the symmetry conditions are imposed about the line $n_2 = 0$ and a linear system is derived which links the displacements along the rows $n_2 = 0$ and $n_2 = 1$. Then, the solvability of this system is considered for various cases of λ, and (4.85) is deduced. The reader is referred to [53, §4] for a detailed discussion and derivation of (4.85). In Figure 4.22, the dispersion relation (4.85) is plotted for several values of r. The in-phase standing wave solution, of the form (4.83), is always given when $\xi = 0$ and corresponds to the minima of the dispersion curves. The frequency of the in-phase standing wave is

$$\omega = \sqrt{\frac{4}{r(2-r)}}, \quad (4.86)$$

whereas for the out-of-phase solution, at $\kappa = \pi$ corresponding to the maxima of the dispersion curves, the frequency is

$$\omega = \sqrt{\frac{2}{r(2-r)} \left[3 + \sqrt{1 + 8(1-r)^2} \right]}. \quad (4.87)$$

4.3.4 From an infinite inclusion to a large finite defect: The case of large N

In this section, it will be demonstrated that the range of eigenfrequencies for which localised eigenmodes exist for the model of finite inclusions described in Section 4.3.1, can be predicted using the model of an infinite chain of defects considered in Section 4.3.3. The motivation for this is as follows. In order to determine the frequencies of localised modes, according to the analysis presented in Section 4.3.1, it is required to solve a transcendental equation (e.g. Equation (4.71)) for ω. Hence, one must resort to numerical methods. Moreover, the equation in question (the solvability condition (4.50)) is obtained by setting the determinant of a matrix to zero. For a system of N defects the matrix is $N \times N$; hence, for a large system of defects, this becomes computationally intensive. However, as will be shown in the current section, if one is merely interested in the range of permitted localised frequencies, this information may be obtained from the dispersion equation of the infinite system.

As an illustrative example, a defect with $N = 20$ particles of non-dimensional mass $r = 0.25$ is embedded within an infinite square lattice. The eigenfrequencies of the finite defect are computed using the method described in Section 4.3.1 and are shown as dash-dot, and dashed, lines in Figure 4.23. In this figure, the eigenfrequency $\omega_{\min} = 3.0374$ corresponds to an in-phase standing wave solution, whereas the frequency $\omega_{\max} = 4.9344$ represents the out-of-phase solution. The maximum and minimum eigenfrequencies are indicated by the dashed lines in Figure 4.23.

Since N is large, it is useful to consider the model of an infinite chain embedded in a square lattice. Expressions (4.86) and (4.87) predict the values of frequency ω for which there exist such solutions. For the numerical values above, the in-phase solution occurs when $\xi = 0$ and $\omega = 3.0237$ and the out-of-phase solution occurs when $\xi = \pi$ and $\omega = 4.9432$. These values of frequency are close to those encountered in the problem of the finite defect for $N = 20$. Moreover, all the eigenfrequencies computed for the finite defect lie within the pass band for the infinite defect, as shown in Figure 4.23.

Figure 4.24 shows the plot of the eigenmodes for the maximum and minimum eigenfrequencies computed for the line defect containing 20 masses. The maximum eigenfrequency $\omega_{\max}$ corresponds to the out-of-phase mode, whereas the minimum eigenfrequency $\omega_{\min}$ gives the in-phase mode.

It is remarked that both the field in Figure 4.24a, and the envelope of the field in Figure 4.24b resemble the first eigenmode of an homogenised rectilinear inclusion. Using this as motivation, the difference operator

$$\mathcal{D}_{\boldsymbol{p}}(\cdot)_{\boldsymbol{p}} = (\cdot)_{\boldsymbol{p}+\boldsymbol{e}_1} + (\cdot)_{\boldsymbol{p}-\boldsymbol{e}_1} + (\cdot)_{\boldsymbol{p}+\boldsymbol{e}_2} + (\cdot)_{\boldsymbol{p}-\boldsymbol{e}_2} - 4(\cdot)_{\boldsymbol{p}}, \qquad (4.88)$$

is introduced. Making use of (4.41), it is found that

$$\left(\frac{\mathcal{D}_{n_1,0}}{\omega^2} + 1\right) u_{n_1,0} = (1-r) \sum_{p=0}^{N-1} u_{p,0} \left(\mathcal{D}_{n_1,0} + \omega^2\right) g(n_1, 0, p; \omega), \qquad (4.89)$$

Dynamic response of elastic lattices and discretised elastic membranes 137

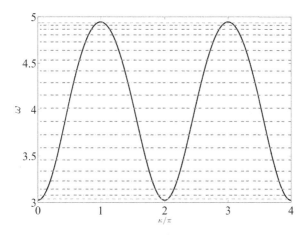

FIGURE 4.23: The dispersion equation (4.85), for the infinite chain, plotted as a function of the normalised Bloch parameter, for $r = 0.25$, represented by the solid curve. Also shown are the blue dash-dot lines corresponding to the eigenfrequencies computed for a finite defect containing $N = 20$ masses. The dashed lines correspond to $\omega_{\min}$ and $\omega_{\max}$.

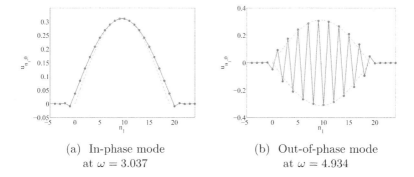

(a) In-phase mode at $\omega = 3.037$

(b) Out-of-phase mode at $\omega = 4.934$

FIGURE 4.24: The solid lines are the eigenmodes for the maximum and minimum eigenfrequencies for a finite line containing 20 defects. The envelope functions defined in (4.94) are shown by the dashed lines.

where $\boldsymbol{n} = (n_1, n_2)$ has been restricted to $\{\boldsymbol{n} : 0 \leq n_1 \leq N-1, n_2 = 0\}$. Since the lattice Green's matrix is a difference kernel (i.e. depends on the difference $|n_1 - p|$),

$$\left(\frac{\mathcal{D}_{n_1,0}}{\omega^2} + 1\right) u_{n_1,0} = (1-r) \sum_{p=0}^{N-1} u_{p,0} \left(\mathcal{D}_{p,0} + \omega^2\right) g(n_1, 0, p; \omega), \quad (4.90)$$

whence, and recalling from (4.39) that $(\mathcal{D}_{\boldsymbol{n}} + \omega^2) g(\boldsymbol{n}, \boldsymbol{p}, \omega) = \delta_{n_1, p} \delta_{n_2, 0}$, it is found that

$$(\mathcal{D}_{\boldsymbol{n}} + r\omega^2) u_{\boldsymbol{n}} = 0, \quad \text{for} \quad \boldsymbol{n} \in \{\boldsymbol{n} : 0 \leq n_1 \leq N-1, n_2 = 0\}. \quad (4.91)$$

It is observed that for a sufficiently large inclusion, the field above and below the inclusion behaves as $u_{n_1,1} = u_{n_1,-1} \approx \lambda u_{n_1,0}$, with $|\lambda| < 1$, in a similar manner to the infinite inclusion. Hence, using (4.91) together with the aforementioned approximation yields

$$u_{n_1+1,0} + u_{n_1-1,0} - 2u_{n,0} + \left[r\omega^2 - 2(1-\lambda)\right] u_{n_1,0} = 0, \quad (4.92)$$

for $0 \leq n_1 \leq N-1$. The first three terms on the left-hand side of (4.92) correspond to the second order central difference operator. Hence, introducing the continuous variable $\eta = n_1$ (where the reader is reminded that the length of the lattice links has been normalised to unity) Equation (4.92) is written as

$$\left[\frac{d^2}{d\eta^2} + r\omega^2 - 2(1-\lambda)\right] u(\eta) = 0. \quad (4.93)$$

The form of Equation (4.93) suggests that the homogenised system is analogous to a string on an elastic foundation, with the constant $2(1-\lambda)$ characterising the effective stiffness of the foundation. It is emphasised that $|\lambda| < 1$ and as such, the stiffness of the elastic foundation is positive.

Consider the problem of an infinite inclusion. According to Equation (4.84), the value of λ corresponding to the lowest eigenmode is $\lambda = 1 - r\omega^2/2$. For this value of λ, the second order derivative vanishes according to Equation (4.93). Moreover, for the displacement at infinity to be finite, $u(\eta)$ must be constant for all η. In this case, the solution of the infinite waveguide problem (4.80) (i.e. $u_{n_1,0} = \text{const}$) is obtained.

For the finite inclusion, it is observed that the displacements at the endpoints are small (see Figure 4.24a). Hence, for a simple estimate it suffices to impose $u(0) = u(N-1) = 0$ whence the solution to (4.93) is

$$u(\eta) = u_0 \sin\left(\eta \sqrt{r\omega^2 - 2(1-\lambda)}\right),$$
$$\text{with } \lambda = 1 + \frac{1}{2}\left[\left(\frac{q\pi}{N-1}\right)^2 - r\omega^2\right], \quad (4.94)$$

where q is an odd number, and u_0 a scaling constant. The first eigenmode corresponds to $\lambda = -0.1396$, which is close to the mean value of λ obtained from the full numerical computation ($\lambda = -0.1426$). The approximation (4.94) for $\lambda = -0.1396$ is shown in Figure 4.24a by the dashed line. The same approximation is used to produce the envelope function shown by the dashed lines in Figure 4.24b. One can observe that this, relatively simple, homogenised model predicts the envelope of the field very well.

4.3.5 Waveguide modes versus waveforms around finite defects

In this section the problem of localised vibrations around a finite rectilinear defect embedded in an infinite square lattice has been discussed in detail. The waveguide problem for an infinite defect has also been briefly described and a comparative analysis of the two classes of problems has been presented.

Although the physical configurations and the methods of analysis of these problems are different, one can observe remarkable properties of solutions, which can be used to make a strong connection. As illustrated in Figure 4.23, the pass band for frequencies of waveguide modes, localised around an infinite chain of masses in a square lattice, contains all eigenmodes describing vibrations localised around a rectilinear defect built of a finite number of masses embedded into the lattice.

In particular, the reader's attention is drawn to the band edges of the dispersion diagram for the infinite defect: Figure 4.23 shows that the frequencies of the eigenmodes for a finite line defect are distributed non-uniformly and they cluster around the edges of the pass band identified for the infinite waveguide. Furthermore the limit, as one approaches the band edge frequency, corresponds to a homogenisation approximation of the linear defect as an inclusion embedded into a homogenised ambient system. The illustrative numerical simulation is produced for an array of 20 masses. It is emphasised that the effect shown is generic and, with an increased number of masses, the density of frequencies of localised modes near the band edges increases.

Symmetric and skew-symmetric modes have been constructed and analysed for a rectilinear "*inclusion*" built of a finite number of masses embedded into the lattice. It has also been shown that the total force exerted on the ambient lattice by the vibrating discrete inclusion is zero for all skew-symmetric modes. Consequently, the displacement fields, associated with skew-symmetric modes, decay at infinity like dipoles, vanishing faster than the displacements corresponding to symmetric modes. This follows from the analytical representations for the solutions and illustrated in Figures 4.17 and 4.18 where the skew-symmetric modes appear to be localised to a much higher degree than symmetric modes. In the aforementioned numerical simulations, the skew-symmetric and symmetric modes appear in pairs, and the frequency of the skew-symmetric mode is higher than the frequency of the corresponding symmetric mode. With reference to Figure 4.15 it is also observed that, in contrast to the one- and three-dimensional multi-atomic cases, there is no lower bound on the perturbation of mass required to initiate a localised mode.

Finally, the reader's attention is drawn to the symmetric and skew-symmetric eigenmodes for a chain of 20 masses shown in Figure 4.24. The corresponding frequencies are the maximum and minimum values in the array of frequencies associated with horizontal lines of Figure 4.23. The envelope curves for both diagrams in Figure 4.24 represent the first eigenmode of a homogenised rectilinear inclusion. The simple homogenised model presented in Section 4.3.4 provides the envelope curves for the finite inclusion. The form of the homogenised system suggests that, macroscopically, the inclusion behaves as a string on an elastic foundation. As expected, the skew-symmetric mode of Figure 4.24b has the higher frequency than the symmetric mode of Figure 4.24a.

Chapter 5

Cloaking and channelling of elastic waves in structured solids

5.1 A cloak is not a shield

There is a significant body of material published on the topic of "cloaking of waves", and the idea of "hiding" a scatterer goes back to 1961 when the paper [58] presented a simple analytical concept, which employed singular maps, and discussed a physical interpretation of transformed equations in the framework of wave propagation in anisotropic inhomogeneous media. This paved the way for an analytical design of a coating, which routes the incoming wave around the scatter, while the scattered field is being suppressed.

At the time of publication, the paper [58] did not receive a large amount of attention from researchers, but thirty years later an elegant paper [136] presented an analytical solution for a problem of the interaction of coated inclusions, in the framework of the multipole method, which had direct implications on modelling of "invisibility cloaks". This article was followed by a seminal contribution [137] addressing physical properties of partially resonant composites. A wide range of physical applications has stimulated further work, leading to a series of papers [28, 49, 76, 98, 120, 121, 123, 141, 143, 155, 156, 158, 174, 175], which have addressed suppression of the scattered fields in problems of wave propagation in media with defects in acoustics, electromagnetism and elasticity.

A shield is understood as a barrier, which protects the interior of an inclusion from an incoming wave. A cloak is a structure encompassing an inclusion, in such a way that the scattered wave is suppressed, i.e. an external observer can detect only the incident wave, and hence the inclusion is "invisible" to an external observer.

There is an intuitive perception that "cloaks" are identical to coatings, which simultaneously shield a defect from an incoming wave and suppress the scattered field. Less emphasis is given to boundary conditions on the interior boundary of the cloaking layers.

We note that the cloak may indeed divert the incoming wave away from the obstacle, but there are also examples which demonstrate that shielding is not always possible, as in [149, 150]. Also, for continuous cloaks, boundary conditions, which are set on the interior boundary of the cloaking layers, deserve attention. In particular, Sections 5.4 and 5.5 address problem of cloaking of flexural waves in plates, and a counter-cloaking example of boundary conditions on the interior boundary of the coating is shown there.

A particular challenge is presented by problems of elasticity, where a singular map, used in the cloaking transformation, distorts the equations of motion of the elastic medium, and a special gauge is required to construct the adequate model, as discussed in the papers [123, 143].

Active cloaks, discussed for problems of flexural waves in elastic plates in [150], show examples when the scattered field can be suppressed by the strategic placement of additional sources, so that, for an external observer, the scatterer becomes "invisible". However resonances may occur leading to high intensity of the physical field in the immediate neighbourhood of the scatterer itself.

The concept highlighted in the present chapter is linked to the boundary condition, specified on the interior boundary of the cloaking coating [27, 45, 48]. This also requires the cloaking transformation to be considered in the framework of the singular perturbation asymptotic procedure [81]. Examples are also considered for scattering of flexural waves, which show scatterers whose cloaking is impossible.

Experimental work on cloaking of flexural waves has been published in [125, 181]. In particular, the paper [125] presents the implementation of the same regularised square cloak that we discuss in the present chapter, and it demonstrates suppression of the scattered flexural wave around a square void in a structured plate.

5.2 Cloaking as a channelling method for incident waves

This section is based on the results of the paper [48], which introduced the novel concept of a non-circular regularised cloak and addressed an important question of the lattice cloaking approximation. A square "push out" transformation, proposed in [168], becomes singular at the inner boundary of the cloaking layer. In contrast, a regularised version of the transformation was used in [48]. This is illustrated in Figure 5.1, where the trapezoids $\chi^{(i)}$ are mapped to the trapezoids $\Omega_-^{(i)}$ with continuity, but not smoothness, imposed on the interfaces between the four trapezoids. The mapping is non-singular on the closure of the cloak, and all corresponding material constants are finite. The regularised transformation yields an effective broadband cloak, with finite material properties, and we also show here that such a cloak may be approximated by a regular lattice.

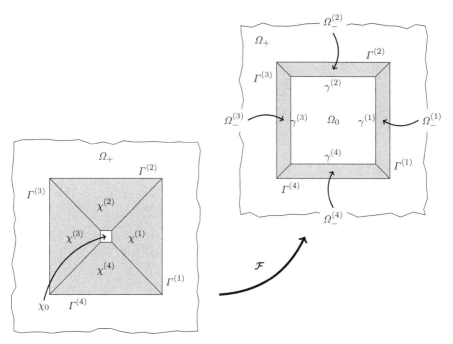

FIGURE 5.1: The map $\mathcal{F}$ maps the undeformed region χ to the deformed configuration Ω_-. The boundary between Ω_+ and $\Omega_-^{(i)}$ is denoted $\Gamma^{(i)}$, while the interface between Ω_0 and $\Omega_-^{(i)}$ is denoted $\gamma^{(i)}$. The corresponding boundaries in the undeformed configuration are denoted by $\Gamma^{(i)}$ and $\sigma^{(i)}$, respectively.

5.2.1 Regularised transformation

Consider a small square $\chi_0 = \{\boldsymbol{X} : |X_1| < \varepsilon, |X_2| < \varepsilon\} \subset \mathbb{R}^2$, which via the map $\mathcal{F}$ is mapped to the square $\Omega_0 = \{\boldsymbol{x} : |x_1| < a, |x_2| < a\} \subset \mathbb{R}^2$. Physically a is the semi-width of the inclusion Ω_0, and ε is the initial semi-width of the square χ_0 where $0 < \varepsilon/a \ll 1$. In this case, it is convenient to decompose the cloak into four sub-domains $\chi = \chi^{(1)} \cup \ldots \cup \chi^{(4)}$, as illustrated in figure 5.1. Formally, the map $\mathcal{F}$ defines a pointwise map from $\boldsymbol{X} \in \chi = \chi^{(1)} \cup \ldots \cup \chi^{(4)}$ to $\boldsymbol{x} = \mathcal{F}(\boldsymbol{X}) \in \Omega_- = \Omega_-^{(1)} \cup \ldots \cup \Omega_-^{(4)}$. The mapping is continuous and non-linear on χ, and defined in a piecewise fashion such that $\mathcal{F} = \mathcal{F}^{(i)}(\boldsymbol{X})$ for $\boldsymbol{X} \in \chi^{(i)}$, where

$$\mathcal{F}^{(1)}(\boldsymbol{X}) = \begin{pmatrix} \alpha_1 X_1 + \alpha_2 \\ \alpha_1 X_2 + \alpha_2 X_2/X_1 \end{pmatrix}, \quad \mathcal{F}^{(2)}(\boldsymbol{X}) = \begin{pmatrix} \alpha_1 X_1 + \alpha_2 X_1/X_2 \\ \alpha_1 X_2 + \alpha_2 \end{pmatrix},$$

$$\mathcal{F}^{(3)}(\boldsymbol{X}) = \begin{pmatrix} \alpha_1 X_1 - \alpha_2 \\ \alpha_1 X_2 - \alpha_2 X_2/X_1 \end{pmatrix}, \quad \mathcal{F}^{(4)}(\boldsymbol{X}) = \begin{pmatrix} \alpha_1 X_1 - \alpha_2 X_1/X_2 \\ \alpha_1 X_2 - \alpha_2 \end{pmatrix},$$

with $\alpha_1 = w/(a + w - \varepsilon)$, $\alpha_2 = (a + w)(a - \varepsilon)/(a + w - \varepsilon)$, and w being the thickness of the cloak. The exterior of the cloak remains unchanged by the map, that is, $\boldsymbol{X} = \boldsymbol{\mathcal{F}}(\boldsymbol{X})$ for $\boldsymbol{X} \in \bar{\Omega}_+$, where the bar denotes the closure of the domain. The Jacobian matrices and determinants are then

$$\boldsymbol{J}^{(1)} = \begin{pmatrix} \alpha_1 & 0 \\ \dfrac{x_2 \alpha_1 \alpha_2}{x_1(\alpha_2 - x_1)} & \dfrac{x_1 \alpha_1}{x_1 - \alpha_2} \end{pmatrix}, \qquad \boldsymbol{J}^{(2)} = \begin{pmatrix} \dfrac{x_2 \alpha_1}{x_2 - \alpha_2} & \dfrac{x_1 \alpha_1 \alpha_2}{x_2(\alpha_2 - x_2)} \\ 0 & \alpha_1 \end{pmatrix},$$

$$\boldsymbol{J}^{(3)} = \begin{pmatrix} \alpha_1 & 0 \\ \dfrac{x_2 \alpha_1 \alpha_2}{x_1(\alpha_2 + x_1)} & \dfrac{x_1 \alpha_1}{x_1 + \alpha_2} \end{pmatrix}, \qquad \boldsymbol{J}^{(4)} = \begin{pmatrix} \dfrac{x_2 \alpha_1}{x_2 + \alpha_2} & \dfrac{x_1 \alpha_1 \alpha_2}{x_2(\alpha_2 + x_2)} \\ 0 & \alpha_1 \end{pmatrix},$$

$$J^{(1)} = \frac{x_1 \alpha_1^2}{x_1 - \alpha_2}, \qquad J^{(2)} = \frac{x_2 \alpha_1^2}{x_2 - \alpha_2},$$

$$J^{(3)} = \frac{x_1 \alpha_1^2}{x_1 + \alpha_2}, \qquad J^{(4)} = \frac{x_2 \alpha_1^2}{x_2 + \alpha_2}.$$

It is emphasised that $J^{(i)}(x_i) = \det \boldsymbol{J}^{(i)} \neq 0$ for $\boldsymbol{x} \in \bar{\Omega}_-^{(i)}$ and $\varepsilon \neq 0$, that is, the map is continuous on both the interior and boundary of the cloak.

Lemma 2.1 in [141] allows the Helmholtz equation for an isotropic homogeneous medium $\mu \nabla_{\boldsymbol{X}} \cdot (\nabla_{\boldsymbol{X}}) u(\boldsymbol{X}) + \varrho \omega^2 u(\boldsymbol{X}) = 0$ for $\boldsymbol{X} \in \chi$ to be written in deformed co-ordinates as

$$[\nabla \cdot (\boldsymbol{C}^{(i)}(\boldsymbol{x}) \nabla) + \rho^{(i)}(\boldsymbol{x}) \omega^2] u(\boldsymbol{x}) = 0, \qquad \boldsymbol{x} \in \Omega_-^{(i)}, \tag{5.1}$$

where μ is the constant ambient stiffness, ϱ is the constant ambient density, $\boldsymbol{C}^{(i)}(\boldsymbol{x}) = [\mu/J^{(i)}(\boldsymbol{x})] \boldsymbol{J}^{(i)}(\boldsymbol{x}) [\boldsymbol{J}^{(i)}(\boldsymbol{x})]^{\mathrm{T}}$ is the transformed stiffness matrix and $\rho^{(i)}(\boldsymbol{x}) = \varrho/J^{(i)}(\boldsymbol{x})$ is the transformed density. The differential operator $\nabla_{\boldsymbol{X}}$ is written in the undeformed space and should be distinguished from ∇ which is written in the deformed coordinates.

Since the mapping is continuous on $\bar{\Omega}_-$, the material properties of the cloak are non-singular. The transformed stiffness tensor is symmetric and positive definite. The density is a scalar and bounded on $\bar{\Omega}_-$. Physically, the transformed material properties correspond to a heterogeneous anisotropic medium.

5.2.1.1 Interface conditions

Without loss of generality, it is convenient to restrict the following analysis to a single side of the cloak. With reference to Figure 5.1, consider the sub-domain $\Omega_-^{(1)} \subset \mathbb{R}^2$ in the absence of the inclusion and remaining three sides of the cloak. In the absence of sources the out-of-plane shear deformation amplitude of an outgoing time-harmonic wave of angular frequency ω satisfies the following equation

$$\mathcal{L} u(\boldsymbol{x}) = 0, \tag{5.2}$$

together with the Sommerfeld radiation condition at infinity. Here
$\mathcal{L} = \nabla \cdot (\boldsymbol{A}(\boldsymbol{x})\nabla) + \rho(\boldsymbol{x})\omega^2$ is the Helmholtz operator, $\boldsymbol{A}(\boldsymbol{x})$ and $\rho(\boldsymbol{x})$ are
defined as

$$\boldsymbol{A}(\boldsymbol{x}) = \begin{cases} \boldsymbol{C}^{(1)}(\boldsymbol{x}) & \text{for } \boldsymbol{x} \in \Omega_-^{(1)} \\ \mu\mathbb{I} & \text{for } \boldsymbol{x} \in \Omega_+ \end{cases}, \quad \rho(\boldsymbol{x}) = \begin{cases} \rho^{(1)}(\boldsymbol{x}) & \text{for } \boldsymbol{x} \in \Omega_-^{(1)} \\ \varrho & \text{for } \boldsymbol{x} \in \Omega_+ \end{cases}.$$
(5.3)

Let $v(\boldsymbol{x})$ be a continuous piecewise smooth solution of the Helmholtz equation in $\mathbb{R}^2$ satisfying the Sommerfeld radiation condition at infinity. Integrating the difference $u(\boldsymbol{x})\mathcal{L}v(\boldsymbol{x}) - v(\boldsymbol{x})\mathcal{L}u(\boldsymbol{x})$ over a disc $\mathcal{D}_r$ of radius r containing $\Omega_-^{(1)}$ yields

$$0 = \int_{\mathcal{D}_r} (u\nabla \cdot \boldsymbol{A}\nabla v - v\nabla \cdot \boldsymbol{A}\nabla u) \, d\boldsymbol{x}$$

$$= \int_{\partial\Omega_-^{(1)}} \left(u^- \boldsymbol{n} \cdot \boldsymbol{A}\nabla v^- - v^- \boldsymbol{n} \cdot \boldsymbol{A}\nabla u^-\right) d\boldsymbol{x}$$

$$- \int_{\partial\Omega_-^{(1)}} \left(u^+ \boldsymbol{n} \cdot \boldsymbol{A}\nabla v^+ - v^+ \boldsymbol{n} \cdot \boldsymbol{A}\nabla u^+\right) d\boldsymbol{x} + \mu \int_{\partial\mathcal{D}_r} (u\boldsymbol{n} \cdot \nabla v + v\boldsymbol{n} \cdot \nabla u) \, d\boldsymbol{x},$$

where the fact that $\nabla u \cdot \boldsymbol{A}\nabla v = \nabla v \cdot \boldsymbol{A}\nabla u$ (since $\boldsymbol{A}$ is symmetric) has already been used. Since $u(\boldsymbol{x})$ and $v(\boldsymbol{x})$ represent outgoing solutions, the final integral vanishes as $r \to \infty$. Thus, the essential interface condition is continuity of the field

$$[u] = 0 \quad \text{on} \quad \partial\Omega_-^{(1)}, \tag{5.4}$$

and the natural interface condition is continuity of tractions

$$\boldsymbol{n} \cdot \boldsymbol{C}^{(1)}\nabla u^- = \mu\boldsymbol{n} \cdot \nabla u^+ \quad \text{on} \quad \partial\Omega_-^{(1)}. \tag{5.5}$$

These interface conditions are equivalent to those derived in [141] for acoustic pressure by application of Nanson's formula and imposing particle continuity.

5.2.1.2 Cloaking problem

Consider the propagation of time harmonic out-of-plane deformations, generated by a point source, in a homogeneous infinite elastic solid in which is embedded an inclusion surrounded by a cloak. The displacement amplitude field then satisfies

$$[\nabla \cdot (\boldsymbol{A}(\boldsymbol{x})\nabla) + \rho(\boldsymbol{x})\omega^2]u(\boldsymbol{x}) = -\delta(\boldsymbol{x} - \boldsymbol{x}_0), \quad \boldsymbol{x} \in \mathbb{R}^2 \setminus \bar{\Omega}_0, \quad \boldsymbol{x}_0 \in \Omega_+ \tag{5.6}$$

$$[\mu_0 \nabla \cdot (\nabla) + \varrho_0\omega^2]u(\boldsymbol{x}) = 0, \quad \boldsymbol{x} \in \Omega_0, \tag{5.7}$$

with continuity of $u(\boldsymbol{x})$ and tractions on all internal boundaries according to (5.4) and (5.5). Additionally, the Sommerfeld radiation condition is imposed at infinity. The stiffness tensor $\boldsymbol{A}(\boldsymbol{x})$ and density $\rho(\boldsymbol{x})$ are

$$\boldsymbol{A}(\boldsymbol{x}) = \begin{cases} \boldsymbol{C}^{(i)}(\boldsymbol{x}) & \text{for } \boldsymbol{x} \in \Omega_-^{(i)} \\ \mu \mathbb{I} & \text{for } \boldsymbol{x} \in \Omega_+ \end{cases}, \qquad \rho(\boldsymbol{x}) = \begin{cases} \rho^{(i)}(\boldsymbol{x}) & \text{for } \boldsymbol{x} \in \Omega_-^{(i)} \\ \varrho & \text{for } \boldsymbol{x} \in \Omega_+ \end{cases}, \qquad (5.8)$$

and μ_0 and ρ_0 are the stiffness and density of the inclusion, respectively.

5.2.1.3 Ray equations

Whilst, in principle, the exact wave behaviour of the displacement field can be obtained from the solution of the cloaking problem, it is illuminating to consider the leading order behaviour of rays through the cloak. Consider a WKB expansion (see [11] and [147] among others) of the displacement amplitude field in terms of angular frequency ω, and the amplitude and phase functions $U_n(\boldsymbol{x})$ and $\varphi(\boldsymbol{x})$, respectively

$$u(\boldsymbol{x}) \sim e^{i\omega\varphi(\boldsymbol{x})} \sum_{n=0}^{\infty} \frac{i^n U_n(\boldsymbol{x})}{\omega^n}, \qquad \text{as } \omega \to \infty, \qquad (5.9)$$

whence the leading order equation for the phase on the interior of the cloak is

$$H(\boldsymbol{x}, \boldsymbol{s}) = 0, \qquad (5.10)$$

where $H(\boldsymbol{x},\boldsymbol{s}) = \mu \varrho^{-1} \boldsymbol{s} \cdot \boldsymbol{g}^{-1} \boldsymbol{s} - 1$, $\boldsymbol{s} = \nabla \varphi$ is the slowness vector, μ and ϱ are the stiffness and density of the ambient medium, respectively, and $\boldsymbol{g}$ is the metric of the transformation. In terms of wave propagation, the conserved quantity $H(\boldsymbol{x},\boldsymbol{s})$ represents the first order slowness contours [134]. The characteristics of the quantity $H(\boldsymbol{x},\boldsymbol{s})$ then satisfy the following system

$$\frac{dH}{dt} = 0, \qquad \frac{d\boldsymbol{x}}{dt} = \frac{\partial H}{\partial \boldsymbol{s}}, \qquad \frac{d\boldsymbol{s}}{dt} = -\frac{\partial H}{\partial \boldsymbol{x}}, \qquad (5.11)$$

where t is the ray parameter (equivalently, the time-like parameter). At this point, it is convenient to introduce index summation notation where summation, from 1 to 2, over repeated indices is implied. The system (5.11) may then be expressed as

$$\frac{ds_i}{dt} = -2\varrho^{-1}\mu s_m s_n J_{nl} \frac{\partial J_{ml}}{\partial x_i}, \qquad \frac{dx_i}{dt} = 2\varrho^{-1}\mu J_{il} J_{jl} s_j, \qquad (5.12)$$

where $J_{ij} = (\boldsymbol{J})_{ij}$ are the components of the Jacobian matrix and should be distinguished from the J, the Jacobian determinant. The superscript labels have been omitted for brevity, but J_{ij} and J should be understood as $J_{ij}^{(k)}$ and $J^{(k)}$ for $k = 1, \ldots 4$ corresponding to the four sides of the cloak.

From Equation (5.10), an alternative representation, in terms of wave normals $\boldsymbol{n}$ and the phase velocity v, is

$$\mu \varrho^{-1} \boldsymbol{n} \cdot \boldsymbol{g}^{-1} \boldsymbol{n} - v^2 = 0. \tag{5.13}$$

The representation (5.13) is obtained by assuming a plane wave solution to the Helmholtz equation (see, for example [134]). Alternatively, seeking a solution of the full wave equation in the form of the leading term in a WKB expansion

$$u(\boldsymbol{x}, t) \sim e^{i\omega \varphi(\boldsymbol{x},t)} \sum_{n=0}^{\infty} \frac{i^n U_n(\boldsymbol{x}, t)}{\omega^n}, \quad \text{as } \omega \to \infty,$$

yields the same result with $\partial \varphi / \partial t = v$. From (5.10) and (5.13) the slowness vector can be expressed in terms of the original material properties (through ϱ and μ) and the map (through $\boldsymbol{J}$) as

$$\boldsymbol{s} = \frac{\boldsymbol{n}}{v} = \frac{\boldsymbol{n}}{|\boldsymbol{J}^T \boldsymbol{n}|} \sqrt{\frac{\varrho}{\mu}}, \tag{5.14}$$

Further, in the undeformed configuration, the equivalent conserved quantities are $\mu \varrho^{-1} \boldsymbol{S} \cdot \boldsymbol{S} - 1 = 0$ and $\mu \varrho^{-1} = V^2$. Together with (5.10) and (5.13), these two equations imply that

$$\boldsymbol{s} = \boldsymbol{J}^{-T} \boldsymbol{S} = \frac{\boldsymbol{J}^{-T} \boldsymbol{N}}{V} = \boldsymbol{J}^{-T} \boldsymbol{N} \sqrt{\frac{\varrho}{\mu}}. \tag{5.15}$$

Now, consider a ray (line) in the ambient medium, in direction $\boldsymbol{N}$ passing through $\boldsymbol{X}_0$ and parameterised by t. The corresponding curve in the cloak is $\boldsymbol{x}(t) = \boldsymbol{\mathcal{F}}(\boldsymbol{X}_0 + t\boldsymbol{N})$, whence

$$\frac{\mathrm{d}x_i}{\mathrm{d}t} = J_{ij} N_j,$$

which using (5.15) may be rewritten thus

$$\frac{\mathrm{d}x_i}{\mathrm{d}t} = J_{il} J_{jl} s_j \sqrt{\frac{\mu}{\varrho}}. \tag{5.16}$$

Taking the derivative of (5.15) for constant $\boldsymbol{N}$ yields

$$\frac{\mathrm{d}s_i}{\mathrm{d}t} = s_k s_n J_{kj} J_{lm} J_{nm} \frac{\partial J^{-1}{}_{ji}}{\partial x_l} \sqrt{\frac{\mu}{\varrho}}.$$

Using the compatibility condition that the deformation gradient should be irrotational under finite deformation $\varepsilon_{jkl} \partial J^{-1}{}_{ik}/\partial x_j = 0_{\ell i}$, the partial derivative above may be written as $\partial J^{-1}{}_{jl}/\partial x_i$, whence

$$\frac{\mathrm{d}s_i}{\mathrm{d}t} = -s_m s_n J_{n\ell} \frac{\partial J_{ml}}{\partial x_i} \sqrt{\frac{\mu}{\varrho}}, \tag{5.17}$$

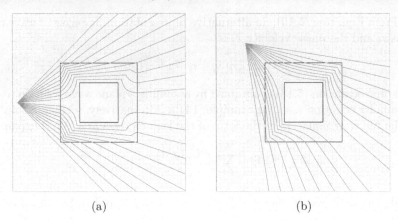

(a)   (b)

FIGURE 5.2: Plots of the ray paths through the cloak for a wave emanated from a point source. Parts (a) and (b) show different positioning of the source, and illustrate that the rays remains unperturbed after passing through the cloaking coating.

where $\varepsilon_{jk\ell}$ is the permutation tensor and the equality $J_{lm}\partial J_{jl}^{-1}/\partial x_i = -J_{jl}^{-1}\partial J_{lm}/\partial x_i$ has already been used. Consider the characteristic equations for the waves in the cloak (5.12), together with the equations of the transformed rays (5.16) and (5.17). The system (5.16) and (5.17) are the equations of characteristics in the cloak, up to an arbitrary scaling constant of $2\sqrt{\mu/\varrho}$. Thus, to leading order, rays (or straight lines) in the ambient medium map directly to rays in the cloak.

Figure 5.2 shows rays emanating from a point source, passing through the cloak and emerging from the cloak on their original trajectory. In this sense, the object is "invisible" to an observer outside the cloaking region. The figure clearly illustrates how wave propagation in the cloak is related to the map.

An interesting alternative perspective is apparent if Figure 5.2 is viewed, not as rays diverging from a source, but as rays converging to a focal point. It is observed that the rays converge to the focal point around the inclusion. One can envisage several applications where such an effect may be useful. For example, image distortion from the mirror mounts in telescopes could be reduced by cloaking the mounts. In addition, apparatus and mounting structures on microwave receivers could be cloaked to improve the quality of the signal. One could also conceive of cloaking mounting points and the surrounding structures in laser cutting machines to protect them from accidental damage.

5.2.1.4 Negative refraction

It is evident from Figure 5.2 that whilst the rays are continuous, they are not necessarily differentiable. In particular, at the interface between the cloak and ambient medium, and at the internal interfaces of the cloak, refraction occurs characterised by a discontinuity of the spatial derivatives of the rays. Of particular interest are the regions on the outer boundary of the cloak where the rays are characterised by a negative index of refraction.

Consider Figure 5.2a. Negative refraction occurs on the right hand interface between the cloak and the ambient medium. A ray exiting the right-hand side of the cloak with gradient M at point $\boldsymbol{X}^{(0)} = \boldsymbol{x}^{(0)}$ may be described by the equation $X_2^{(s)} - X_2^{(0)} = M(X_1^{(s)} - X_1^{(0)})$ in the ambient medium, where $\boldsymbol{X}^{(s)}$ is the position of the source. The behaviour of the ray at the interface is entirely characterised by the position of the source relative to the interface and its initial gradient. Therefore, without loss of generality, the following analysis may be restricted to the right hand side of the cloak. On the interior of the right-hand side of the cloak, the ray exiting the cloak at $\boldsymbol{x}^{(0)}$ is characterised by

$$x_2 = x_1 \left(M + \frac{\alpha_1(x_2^{(0)} - Mx_1^{(0)})}{x_1 - \alpha_2} \right).$$

The gradient of the ray as it approaches the exterior boundary from the interior of the cloak is then

$$m^* = \lim_{x_1 \to (a+w)^-} \frac{\mathrm{d}x_2}{\mathrm{d}x_1} = \frac{M(a+w-\varepsilon)(a+w) - x_2^{(0)}(a+\varepsilon)}{w(a+w)}.$$

Thus, the gradient is discontinuous at the exterior interface. For negative refraction it is required that $m^*M < 0$, which leads to the following inequalities

$$0 < M < x_2^{(0)} \frac{a+\varepsilon}{(a+w)(a+w-\varepsilon)}, \text{ or } x_2^{(0)} \frac{a+\varepsilon}{(a+w)(a+w-\varepsilon)} < M < 0. \tag{5.18}$$

Assuming the source lies on the line $X_2 = 0$ as in Figure 5.2a the inequalities reduce to the single inequality

$$X_1^{(s)} < -\frac{(a+w)(w-2\varepsilon)}{a+\varepsilon},$$

which is satisfied for all sources outside the cloak $X_1^{(s)} < -(a+w)$, if $w < a+3\varepsilon$. Thus, for a sufficiently thin cloak and a cylindrical source placed along $X_2 = 0$ and at any distance from the cloak, negative refraction is expected on the opposite side of cloak.

Alternatively, for a source located along the line $X_1 = 0$ the inequalities (5.18) become

$$0 < X_2^{(s)} < \frac{(a+w)(w-2\varepsilon)}{(a+w-\varepsilon)}, \text{ or } -\frac{(a+w)(w-2\varepsilon)}{(a+w-\varepsilon)} < X_2^{(s)} < 0, \tag{5.19}$$

where the fact that $|x_2^{(0)}| < (a+w)$ has been used. Since $a, w > 0$, and $0 < \varepsilon/a \ll 1$, the above inequalities are never satisfied. Hence, the lack of negative refraction on the horizontal interfaces in Figure 5.2a. Similar arguments may be used to consider other regions where negative refraction may, or may not occur. It is observed that negative refraction always occurs at the interface between the four regions of the cloak, where the material properties (equivalently the transformation) are not smooth.

5.2.1.5 Scattering measure

It is desirable to have some quantifiable measure of the quality of the cloak, rather than relying on visual observations. However, it is not obvious what "quality" means with respect to a cloak, given that there are essentially three fields involved, i.e. the incident wave in the absence of both cloak and inclusion, the uncloaked field with an inclusion present but without a cloak, and the cloaked field with both the inclusion and cloak. Previous experimental works [181] have used an L_2 norm computed directly from the measured fields to place a numerical value on the quality of the cloak. It is in this spirit that the following "*scattering measure*" is formally introduced as a tool to quantify the cloaking effect

$$\mathcal{E}(u_1, u_2, \mathcal{R}) = \left(\int_\mathcal{R} |u_1(\boldsymbol{x}) - u_2(\boldsymbol{x})|^2 \, \mathrm{d}\boldsymbol{x} \right) \left(\int_\mathcal{R} |u_2(\boldsymbol{x})|^2 \, \mathrm{d}\boldsymbol{x} \right)^{-1}, \qquad (5.20)$$

where $\mathcal{R} \subset \mathbb{R}^2$ is some region outside the cloak, and $u_1(\boldsymbol{x})$ and $u_2(\boldsymbol{x})$ are any two fields. The quantities $\mathcal{E}(u_u, u_0, \mathcal{R})$ and $\mathcal{E}(u_c, u_0, \mathcal{R})$ are presented for a series of illustrative simulations. The field $u_0(\boldsymbol{x}) = i\,\mathrm{H}_0^{(1)}(\omega\sqrt{\rho/\mu}|\boldsymbol{x} - \boldsymbol{x}_0|)/4$ is the Green's function for the unperturbed problem and represents the "ideal" field, $u_u(\boldsymbol{x})$ and $u_c(\boldsymbol{x})$ are the uncloaked and cloaked fields, respectively. Thus, perfect cloaking corresponds to a vanishing $\mathcal{E}$. Along with the raw scattering measures an additional quantity, $Q = |\mathcal{E}(u_u, u_0, \mathcal{R}) - \mathcal{E}(u_c, u_0, \mathcal{R})|/\mathcal{E}(u_u, u_0, \mathcal{R})$, is also presented. The parameter Q characterises the relative reduction of the scattering measure by the introduction of a cloak. It should be emphasised that this is only one of a number of possible measures of quality.

5.2.1.6 Choice of $\mathcal{R}$

For the purpose of illustration three different regions of integration are considered, as shown in Figure 5.3. The three regions were chosen as follows: (a) $\mathcal{R}_1$ is the region taking into account significant near field effects and a wide range of scattering angles. (b) The forward scattering region ($\mathcal{R}_2$) is relevant if the scattered field is measurable over a wide range of forward scattering angles. (c) The corner scattering region ($\mathcal{R}_3$) is employed for sources located along the diagonal of the square inclusion. It is emphasised that $\|\mathcal{R}_1\| \neq \|\mathcal{R}_2\| = \|\mathcal{R}_3\|$, and the leading edges of the regions $\mathcal{R}_2$ and $\mathcal{R}_3$ are located at the same distance from the source.

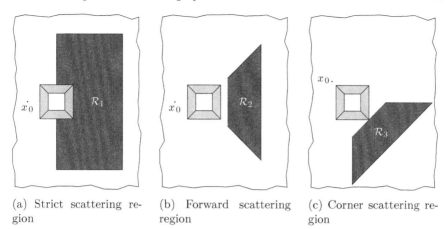

(a) Strict scattering region

(b) Forward scattering region

(c) Corner scattering region

FIGURE 5.3: The three regions used for computation of the scattering measure.

5.2.1.7 Illustrative simulations

Here we show several illustrative examples of a regularised cloak suppressing the scattered field for a wave emanating from a point source. The field plots, presented in Figures 5.4 and 5.5, were created using the finite element software COMSOL Multiphysics®. Perfectly matched layers were used in the vicinity of the boundary of the computational domain in order to simulate an infinite domain. For the purposes of these computations, the following non-dimensional parameter values* were chosen: $a = 0.5$, $w = 0.5$, $\mu = \varrho = 1$, $\mu_0 = 0.1$, $\varrho_0 = 0$, $\varepsilon = 1 \times 10^{-6}$. Figures 5.4 and 5.5 show the displacement field $u(\boldsymbol{x})$ for a cylindrical source oscillating at $\omega = 5$ and $\omega = 10$, respectively. The figures clearly illustrate the efficacy of the square cloak, even at relatively high frequencies. Table 5.1 shows the corresponding scattering measures as introduced in Section 5.2.1.5. It is clear that this square *"push out"* cloak is highly effective. Indeed, for the illustrative simulations presented here, the cloak reduces the scattering measure by not less than 99.62% compared with the uncloaked inclusion.

Figure 5.6 shows the scattering measure plotted against the angular frequency ω, normalised to the case when $\mu = \varrho = 1$. The solid curve in Figure 5.6 corresponds to the continuum, in the absence of both cloak and inclusion. This curve gives an indication of the numerical error in the simulation induced by, for example, the use of perfectly matched layers and the numerical discretisation. The dashed curve corresponds to the cloaked inclusion, whilst the dash-dot curve corresponds to the uncloaked inclusion. It is observed that the numerical measure of the cloaked inclusion remains close to that of the intact

---

*Throughout this section, all numerical parameters are normalised such that $\mu = \varrho = 1$ unless otherwise stated.

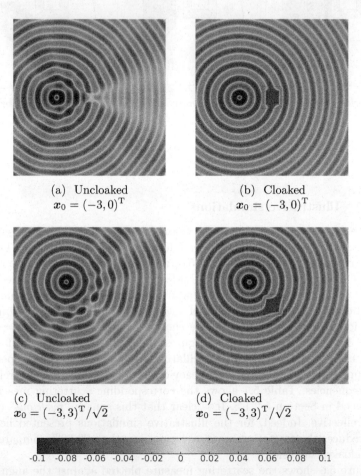

FIGURE 5.4: Plots of the field u for the uncloaked and cloaked square inclusion, where the angular frequency of excitation is $\omega = 5$. The position of the source is indicated under the relevant plot.

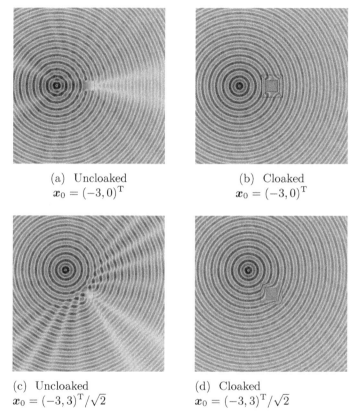

(a) Uncloaked
$x_0 = (-3, 0)^T$

(b) Cloaked
$x_0 = (-3, 0)^T$

(c) Uncloaked
$x_0 = (-3, 3)^T/\sqrt{2}$

(d) Cloaked
$x_0 = (-3, 3)^T/\sqrt{2}$

FIGURE 5.5: Plots of the field u for the uncloaked and cloaked square inclusion where the angular frequency of excitation is $\omega = 10$. The position of the source is indicated under the relevant plot and the inclusion is located at the centre of the image in all cases.

TABLE 5.1: The scattering measures corresponding to the simulations shown in Figures 5.4 and 5.5.

Source Position	Frequency	Scattering Measure $\mathcal{E}$ Uncloaked	Cloaked	Q
Scattering region $\mathcal{R}_1$				
$(-3, 0)^\mathrm{T}$	5	0.1529	4.351×10^{-4}	0.9972
$(-3, 0)^\mathrm{T}$	10	0.1455	4.514×10^{-4}	0.9969
$(-3, 3)^\mathrm{T}/\sqrt{2}$	5	0.2002	3.941×10^{-4}	0.9980
$(-3, 3)^\mathrm{T}/\sqrt{2}$	10	0.3286	4.068×10^{-4}	0.9988
Scattering region $\mathcal{R}_2$				
$(-3, 0)^\mathrm{T}$	5	0.3224	3.664×10^{-4}	0.9989
$(-3, 0)^\mathrm{T}$	10	0.3093	1.167×10^{-3}	0.9962
Scattering region $\mathcal{R}_3$				
$(-3, 3)^\mathrm{T}/\sqrt{2}$	5	0.2988	3.654×10^{-4}	0.9988
$(-3, 3)^\mathrm{T}/\sqrt{2}$	10	0.2988	7.803×10^{-4}	0.9974

continuum for a large range of frequencies. Moving to dimensional quantities, suppose the simulation corresponded to a particular polarisation of an electric wave travelling through glass at a speed of approximately 2×10^8 m·s^{-1}. The line $\omega = 10$ on Figure 5.6 then corresponds to a frequency of approximately 340 MHz. We also emphasise on the broadband feature of the approximate cloak described above, which makes it highly attractive with wider application compared to many other designs.

5.2.2 Cloaking path information

In recent years there has been much interest in experiments to elucidate the fundamental principles of quantum mechanics, and in particular the relationship between measurement and system behaviour. One basic experiment which with its variants features in many such experimental studies is the classical Young's double slit experiment (see, for example, [80]). This suggested that it may be of interest to consider the interaction of the excellent mechanical cloaking demonstrated earlier with the foundational quantum mechanics experiment.

A Young's double slit experiment, presented for the square cloak in [48], is discussed here for the case when a monochromatic plane wave is incident on a rigid screen with two apertures. Due to the superposition of the waves passing through the two apertures, the distinctive double slit interference pattern is produced on an observation screen placed on the opposite side of the apertures to the source. The result of a simulation of the standard experiment is shown in Figure 5.7a, with the diffraction pattern produced on the observation screen

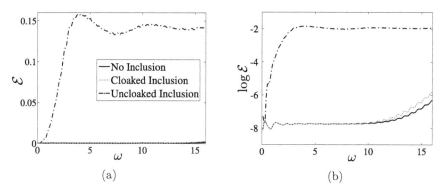

FIGURE 5.6: (a) The figure shows the scattering measure plotted against angular frequency. (b) The figure shows the log of the scattering measure plotted against angular frequency. The solid line corresponds to the continuum in the absence of both an inclusion and cloak. The dashed line represents the cloaked inclusion and the dash-dot line corresponds to the uncloaked inclusion. The region $\mathcal{R}_1$, shown in Figure 5.3, was used to compute the error measure.

(in this case, a vertical line near the right hand edge of Figures 5.7a–5.7c) shown as curve (a) in Figure 5.7d. Placing an object (inclusion) near to one slit, as in Figure 5.7b, partially destroys the diffraction pattern. The corresponding pattern on the observation screen is shown as line (b) in Figure 5.7d. However, coating the object with the square *push out* cloak presented earlier, as shown in Figure 5.7c, restores the original diffraction pattern almost entirely. The interference pattern corresponding to the cloaked object is shown as curve (c) in Figure 5.7d.

It has been conclusively demonstrated that the cloaking is of sufficient quality to render the interference pattern almost immune to movement in the position of the cloaked obstacle. In particular, movement of the cloaked obstacle, it would seem, does not yield any information about the passage of waves through one slit or the other. This consideration would be important if one were able to carry out an experiment in which single quantised elements of flexural vibration were in the system at any given instance in time. The quantum mechanical view would be that, if no path information were available from measurements, the interference fringes behind the double slit should persist.

5.2.3 Cloaking with a lattice

Cloaks designed using transformation optics may have such extreme physical attributes that the requisite materials cannot be physically realised without recourse to metamaterials. It is with this motivation in mind that the following approximate cloak in the low frequency regime is developed. The cloak

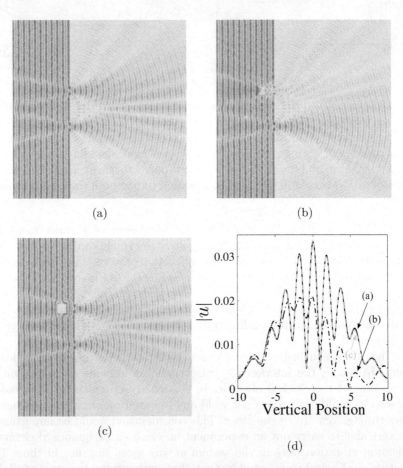

FIGURE 5.7: (a)–(c) The field $u(\boldsymbol{x})$ for the Young's double slit experiment with no inclusion, an uncloaked inclusion, and a cloaked inclusion respectively. (d) A plot of $|u(\boldsymbol{x})|$ over the observation screen illustrating the interference fringes for cases (a)–(c).

Cloaking and channelling of elastic waves in structured solids   157

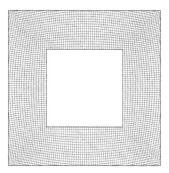

FIGURE 5.8: The lattice formed from the principal directions of the stiffness matrix for the continuum cloak.

is constructed as an approximation to the continuum square cloak considered earlier, but is realised using a discrete lattice structure, formed from rods and point masses. The advantage of a discrete structure over a continuous material is that much higher contrasts in material properties are easily realisable using lattices. The development of an approximate cloaking material using a lattice may allow the practical construction of cloaks. In the following discussion, it is emphasised that repeated indices do not imply summation.

With reference to the formulae for the Jacobian of the transformation in Section 5.2.1, the symmetric stiffness matrices $\boldsymbol{C}^{(i)} = [\mu/J^{(i)}](\boldsymbol{J}^{(i)})^{\mathrm{T}}\boldsymbol{J}^{(i)}$ are positive definite. Therefore, the stiffness matrix admits the following diagonalisation

$$\boldsymbol{C}^{(i)} = \boldsymbol{P}^{(i)\mathrm{T}}\boldsymbol{\Lambda}^{(i)}\boldsymbol{P}^{(i)}, \tag{5.21}$$

where $\boldsymbol{P}^{(i)} = (\boldsymbol{e}_1^{(i)}, \boldsymbol{e}_2^{(i)})$ are the matrices with columns consisting of the principal directions (eigenvectors) of $\boldsymbol{C}^{(i)}$, and $\boldsymbol{\Lambda}^{(i)} = \mathrm{diag}(\lambda_1^{(i)}, \lambda_2^{(i)})$ is the diagonal matrix of the corresponding ordered [positive] eigenvalues such that $\lambda_1^{(i)} > \lambda_2^{(i)}$. The eigenvectors yield the principal lattice vectors of the locally orthogonal lattice with homogenised stiffnesses $\lambda_j^{(i)}$ in direction $\boldsymbol{e}_j^{(i)}$. In particular, the lattice nodes lie at the intersection points of the solutions of the following non-linear system of first order differential equations

$$\frac{\mathrm{d}}{\mathrm{d}\tau}\boldsymbol{x}_j^{(i)} = \boldsymbol{e}_j^{(i)}(\boldsymbol{x}_j^{(i)}), \qquad \text{for } i = 1,\ldots 4, \text{ and } j = 1, 2, \tag{5.22}$$

for some array of initial positions, where $\boldsymbol{x}_j^{(i)}$ is the position vector along the characteristic defined by $\boldsymbol{e}_j^{(i)}$ inside the ith side of the cloak and τ parametrises the curve. Naturally, this would lead to a lattice with curved links. However, for a sufficiently refined lattice the curved members may be replaced with linear links. The lattice links are then the linearisation of the characteristic between two neighbouring nodes on the characteristic. Figure 5.8 shows the geometry of the lattice formed from the principal vectors of the stiffness matrix. Requiring local conservation of flux allows the stiffness of the lattice link parallel to $e_j^{(i)}$ to be determined as $\ell_{ij}\lambda_j^{(i)}$, where ℓ_{ij} is the length of the link along $e_j^{(i)}$. The distribution of nodal mass may be determined by evaluating the integral

$$m(\boldsymbol{x_p}) = \int_{\mathcal{A}(\boldsymbol{x_p})} \rho(\boldsymbol{x})\,\mathrm{d}\boldsymbol{x},$$

over the unit cell $\mathcal{A}(\boldsymbol{x_p})$ containing the lattice node at $\boldsymbol{x_p}$.

In principle, the lattice cloak may be constructed exactly as described above and illustrated in Figure 5.8. However, for narrow cloaks where $w/a \ll 1$, the locally orthogonal lattice depicted in Figure 5.8 may be approximated by a globally orthogonal regular square lattice. A regular square lattice is more convenient to implement compared with the non-globally orthogonal lattice generated from the spectral decomposition of the stiffness matrix. Although the geometry of the approximate lattice is regular, it should be emphasised that the stiffness of the links and mass of the nodes vary with position according to the projection of $\boldsymbol{A}(\boldsymbol{x})$ and $\rho(\boldsymbol{x})$ as described above.

5.2.3.1 Geometry and governing equations for an inclusion cloaked by a globally orthogonal lattice

Consider a square inclusion $\Omega_0 = \{\boldsymbol{x} : |x_1| < a, |x_2| < a\}$, $a > 0$, embedded in $\mathbb{R}^2$, surrounded by a cloak $\Omega_- = \{\boldsymbol{x} : a < |x_1| < a+w, a < |x_2| < a+w\}\backslash\Omega_0$, where $w > 0$ is the thickness of the cloak. The cloak consists of a discrete lattice structure with lattice points at $\boldsymbol{x} = \ell\boldsymbol{p}$, where $\boldsymbol{p} \in \mathbb{Z}^2 \cap \{\boldsymbol{n} : \ell\boldsymbol{n} \in \Omega_-\}$. The lattice is statically anisotropic with links parallel and perpendicular to the boundaries having contrasting material properties, as shown in Figure 5.9.

As for the continuum cloak, solutions of the Helmholtz equation are of primary interest. In particular, the following problem for the field $u(\boldsymbol{x})$ is studied

$$[\mu\nabla \cdot (\nabla) + \varrho\omega^2]u(\boldsymbol{x}) = -\delta(\boldsymbol{x} - \boldsymbol{x}_0), \quad \boldsymbol{x}, \boldsymbol{x}_0 \in \Omega_+, \quad (5.23)$$

$$[\mu_0\nabla \cdot (\nabla) + \varrho_0\omega^2]u(\boldsymbol{x}) = 0, \quad \boldsymbol{x} \in \Omega_0, \quad (5.24)$$

$$m(\boldsymbol{p})\omega^2 u(\boldsymbol{p}) + \sum_{\boldsymbol{q}\in\mathcal{N}(\boldsymbol{p})} \ell\eta(\boldsymbol{q},\boldsymbol{p})\left[u(\boldsymbol{p}+\boldsymbol{q}) - u(\boldsymbol{p})\right] = 0, \quad \text{in } \Omega_-, \quad (5.25)$$

where $e_i = (\delta_{i1}, \delta_{i2})^\mathrm{T}$, $\boldsymbol{p} \in \mathbb{Z}^2$, and $\mathcal{N} = \{\pm e_1, \pm e_2\}$ is the set of nearest neighbours. The stiffness of the lattice links are the restriction of the eigenvalues of the stiffness matrix to the links. In particular, for the link connecting nodes $\boldsymbol{p}$ and $\boldsymbol{p}+\boldsymbol{q}$, $\eta(\boldsymbol{q},\boldsymbol{p})$ takes the value $\lambda_1^{(i)}|_{[\ell\boldsymbol{p},\ell(\boldsymbol{p}+\boldsymbol{q})]}$ if the vector $\boldsymbol{q}$ is parallel to the exterior boundary of the cloak, $\Gamma^{(i)}$, and $\lambda_2^{(i)}|_{[\ell\boldsymbol{p},\ell(\boldsymbol{p}+\boldsymbol{q})]}$ otherwise. The corner regions are matched as illustrated in Figure 5.9. Here, $\lambda_j^{(i)}|_{[\ell\boldsymbol{p},\ell(\boldsymbol{p}+\boldsymbol{q})]}$ indicates the restriction of $\lambda_j^{(i)}$ to the line $[\ell\boldsymbol{p}, \ell(\boldsymbol{p}+\boldsymbol{q})]$. The associated interface

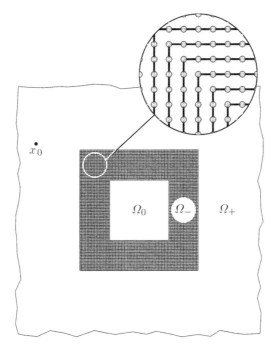

FIGURE 5.9: The lattice cloak Ω_-, surrounding the square inclusion Ω_0, embedded in the ambient medium Ω_+. The thick black lines in the lattice cloak indicate links of high stiffness, while the grey lines indicate links of low stiffness.

conditions corresponding to continuity of tractions are

$$\boldsymbol{n} \cdot \nabla u(\boldsymbol{x}) = \begin{cases} 0 & \text{for } \boldsymbol{x} \in \partial\Omega^- \text{ and } \boldsymbol{x} \pm \ell\boldsymbol{q} \notin \Omega_- \\ \ell\eta(\mp\boldsymbol{q},\boldsymbol{p})u(\boldsymbol{x} \pm \ell\boldsymbol{q})/\mu & \text{for } \boldsymbol{x} \in \bigcup_i \Gamma^{(i)} \text{ and } \boldsymbol{x} \pm \boldsymbol{q} \in \Omega_- \\ \ell\eta(\mp\boldsymbol{q},\boldsymbol{p})u(\boldsymbol{x} \pm \ell\boldsymbol{q})/\mu_0 & \text{for } \boldsymbol{x} \in \bigcup_i \gamma^{(i)} \text{ and } \boldsymbol{x} \pm \boldsymbol{q} \in \Omega_- \end{cases}$$
(5.26)

where $i = 1, \ldots, 4$, and the Sommerfeld radiation condition at infinity. The quantity $\eta(\boldsymbol{q},\boldsymbol{p})$ is the projection of the diagonalised stiffness matrix onto the lattice link connecting lattice points $\boldsymbol{p}$ and $\boldsymbol{p} + \boldsymbol{q}$.

Physically, (5.23)–(5.26) corresponds to the problem of the propagation of time-harmonic waves of angular frequency ω generated by a point load at $\boldsymbol{x}_0$. The field $u(\boldsymbol{x})$ then corresponds to the out-of-plane displacement amplitude and ϱ is the scalar density. The region Ω_- consists of an array of nodes of mass m, connected by massless rods of length ℓ and stiffness according to their orientation.

5.2.3.2 Illustrative lattice simulations

The approximate lattice cloaks were examined using the finite element software COMSOL Multiphysics®. Perfectly matched layers were used in the vicinity of the boundary of the computational domain in order to simulate an infinite domain. For the purpose of illustration, a square of semi-width $a = 0.5$, surrounded by a lattice cloak with $w = 0.1$ and links of length 5×10^{-3} was used. The inclusion is located at the origin of the computational window.

5.2.3.3 Basic lattice cloak

Before proceeding to the illustrative simulations for the regular lattice with heterogeneous distributions of stiffness and mass, it is instructive to consider a simple approximation. Many cloaks created from transformation optics have the general characteristic of having a high phase speed parallel to the boundary of the cloak, and a low phase speed in the direction normal to the boundary (see [57] among others). Therefore, as an initial approximation, the case of a regular square lattice with a homogeneous, but orthotropic distribution of stiffness and a homogeneous distribution of mass is considered. Consider the right-hand side of the cloak $\Omega_-^{(1)}$. For a narrow cloak with $w/a \ll 1$, $x_1 \sim a+w$ and hence the density may be approximated by $\rho \sim 1 + a/w$. The greatest contrast in stiffness occurs at $x_2 = 0$, thus the vertical links are assigned stiffness $\ell\lambda_1^{(1)}(a+w,0)$ and the horizontal links stiffness $\ell\lambda_2^{(1)}(a+w,0)$. The mass of the nodes is $\ell^2(1 + a/w)$. The material properties of the remaining three sides of the cloak are adjusted accordingly.

Figures 5.10 and 5.11 show the field $u(\boldsymbol{x})$ for the uncloaked inclusion (a) and (d), and the inclusion cloaked with this *basic* cloak (b) and (e). For $\omega = 3$ Figure 5.10 indicates that the *basic* cloak partially mitigates the shadow cast by the inclusion and acts to reform the cylindrical wave fronts behind the inclusion. As illustrated by Figure 5.11, this partial cloaking effect deteriorates

TABLE 5.2: The scattering measures corresponding to the simulations for the *basic lattice model* shown in Figures 5.10 and 5.11.

Source Position	Frequency	Scattering Measure $\mathcal{E}$ Uncloaked	Cloaked	Q
		Scattering region $\mathcal{R}_1$		
$(-3, 0)^{\text{T}}$	3	0.1430	0.1662	0.1617
$(-3, 3)^{\text{T}}/\sqrt{2}$	3	0.1113	0.1816	0.6327
$(-3, 0)^{\text{T}}$	5	0.1529	0.2495	0.6318
$(-3, 3)^{\text{T}}/\sqrt{2}$	5	0.2002	0.3538	0.7676
		Scattering region $\mathcal{R}_2$		
$(-3, 0)^{\text{T}}$	3	0.2341	0.3362	0.4363
$(-3, 0)^{\text{T}}$	5	0.3224	0.4671	0.4489
		Scattering region $\mathcal{R}_3$		
$(-3, 3)^{\text{T}}/\sqrt{2}$	3	0.1578	0.3455	1.189
$(-3, 3)^{\text{T}}/\sqrt{2}$	5	0.2988	0.6011	1.012

with increasing frequency. Indeed, in some cases, the presence of the lattice cloak seems to increase the shadow region. Table 5.2 details the values of the scattering measures for the fields illustrated in Figures 5.10 and 5.11. The scattering measures shown in Table 5.2 suggest that, although visually the basic lattice cloak appears to work reasonably well, this may not be the case. The fact that the basic lattice cloak increases the scattering measure compared with the uncloaked inclusion further emphasises the need for an objective measure of the quality of cloaks, rather than simply relying on visual observations.

This increase in the scattering measure by the basic lattice cloak motivates the introduction of the following refined model.

5.2.3.4 Refined lattice cloak

Consider now the lattice described in Section 5.2.3.1, i.e. the regular square lattice with inhomogeneous distribution of stiffness and mass. Figures 5.10 and 5.11 show the field $u(\boldsymbol{x})$ for the uncloaked inclusion and the inclusion with a lattice cloaking. With reference to the simulations for the *basic* cloak (b) and (e) the *refined* lattice cloak (c) and (f), it is observed that the efficiency of the *refined* lattice cloak, whilst not as high as that of the continuum cloak, is much greater than that of the *basic* cloak. The table of scattering measures for the approximate cloak is shown in Table 5.3 and further evidences the effectiveness of the *refined* lattice cloak. Indeed, for several simulations (in particular those where the scattering measure is taken over the forward or corner scattering regions $\mathcal{R}_1$ and $\mathcal{R}_2$, respectively) the efficiency of the *refined*

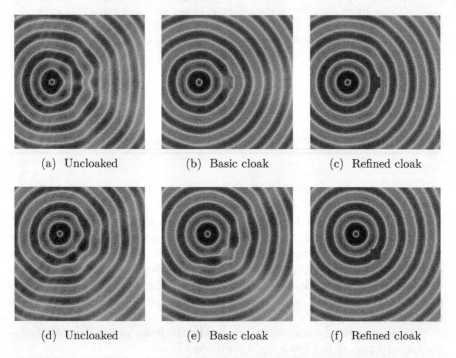

FIGURE 5.10: Plots of the field $u(\boldsymbol{x})$ for a cylindrical wave incident on a square inclusion in the absence of a cloak (parts (a) and (d)), a square inclusion coated with the *basic* lattice (parts (b) and (e)), and an inclusion coating with the refined lattice (parts (c) and (f)). Here the angular frequency of excitation is $\omega = 3$ and the source is located at $\boldsymbol{x}_0 = (-3, 0)^{\mathrm{T}}$ in (a)–(c), and at $\boldsymbol{x}_0 = (-3, 3)^{\mathrm{T}}/\sqrt{2}$ in (e)–(f).

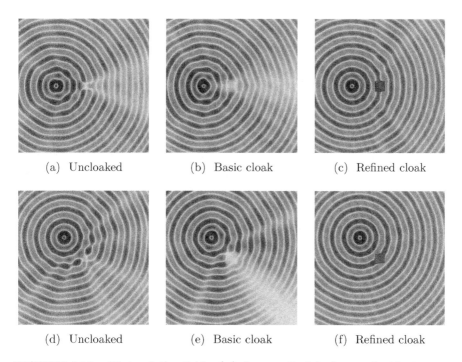

FIGURE 5.11: Plots of the field $u(x)$ for a cylindrical wave incident on a square inclusion in the absence of a cloak (parts (a) and (d)), a square inclusion coated with the *basic lattice model* (parts (b) and (e)), and an inclusion coating with the refined lattice (parts (c) and (f)). Here the angular frequency of excitation is $\omega = 10$ and the source is located at $x_0 = (-3, 0)^T$ in (a)–(c), and at $x_0 = (-3, 3)^T/\sqrt{2}$ in (e)–(f).

TABLE 5.3: The scattering measures corresponding to the simulations shown for the *refined lattice model* in Figures 5.10 and 5.11.

Source Position	Frequency	Scattering Measure $\mathcal{E}$ Uncloaked	Cloaked	Q
\multicolumn{5}{c}{Scattering region $\mathcal{R}_1$}				
$(-3,0)^T$	3	0.1430	0.01191	0.8929
$(-3,3)^T/\sqrt{2}$	3	0.1113	3.385×10^{-3}	0.9763
$(-3,0)^T$	5	0.1529	0.04324	0.7173
$(-3,3)^T/\sqrt{2}$	5	0.2002	0.03125	0.8438
\multicolumn{5}{c}{Scattering region $\mathcal{R}_2$}				
$(-3,0)^T$	3	0.2341	0.01150	0.9508
$(-3,0)^T$	5	0.3224	0.0172	0.9508
\multicolumn{5}{c}{Scattering region $\mathcal{R}_3$}				
$(-3,3)^T/\sqrt{2}$	3	0.1578	5.047×10^{-3}	0.9680
$(-3,3)^T/\sqrt{2}$	5	0.2988	0.02114	0.9292

cloak in reducing the scattering measure approaches that of the continuum cloak.

As expected the effectiveness of the lattice cloaks reduce with increasing frequency. However, for sufficiently low frequencies the *refined* lattice cloak in particular, works well.

5.3 Boundary conditions on the interior contour of a cloak

Whilst cloaking via transformation geometry has been extensively treated in the literature, the sensitivity of the cloaking effect to the boundary conditions is rarely discussed. The cloak is formed by deforming a small region (a point in the case of the classical radial transformation [158]), into a larger finite region. If the region is an inclusion, then the natural interface conditions may be determined following the method outlined in Sections 5.2.1.1 and 5.2.1.2. If the cloaked region is a void or rigid inclusion, however, there is some freedom in choosing the boundary condition, subject to the constraints of the physical problem. Figure 5.12 shows the field $u(\boldsymbol{x})$ for a cloaked void, with Neumann (parts (a) and (b)) and Dirichlet (parts (c) and (d)) conditions applied to the interior of the cloaked region. The corresponding scattering measures are shown in Table 5.4.

Although the square cloak is effective in both cases, it is clear from both

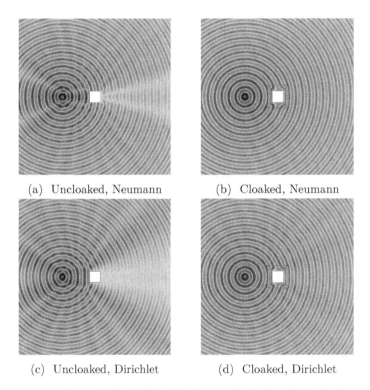

(a) Uncloaked, Neumann (b) Cloaked, Neumann

(c) Uncloaked, Dirichlet (d) Cloaked, Dirichlet

FIGURE 5.12: Plots of the field u for the uncloaked and cloaked square inclusion with Neumann boundary conditions in parts (a) and (b), and Dirichlet boundary conditions in parts (c) and (d). Here the source is located at $\boldsymbol{x} = (-3, 0)^{\mathrm{T}}$ and oscillates at $\omega = 10$.

TABLE 5.4: The scattering measures for a void with Neumann and Dirichlet boundary conditions. Here the source is located at $[-3,0]^T$.

Source Boundary Condition	Frequency	Scattering Measure $\mathcal{E}$ Uncloaked	Cloaked	Q
		Scattering region $\mathcal{R}_1$		
Neumann	5	0.1624	4.351×10^{-4}	0.9973
Neumann	10	0.1558	4.540×10^{-4}	0.9971
Dirichlet	5	0.2931	1.038×10^{-2}	0.9646
Dirichlet	10	0.2553	7.875×10^{-3}	0.9692
		Scattering region $\mathcal{R}_2$		
Neumann	5	0.3436	3.664×10^{-4}	0.9989
Neumann	10	0.3258	1.163×10^{-3}	0.9964
Dirichlet	5	0.4864	1.566×10^{-2}	0.9678
Dirichlet	10	0.5030	1.673×10^{-2}	0.9667

the figures and the table of scattering measures that the type of boundary condition imposed on the cloaked object affects the quality of the cloaking. Indeed, for a void (Neumann) the cloaking reduces the scattering measure by between 99.7% and 99.9% for both $\omega = 5$ and $\omega = 10$. In contrast, cloaking reduces the scattering measure of a rigid inclusion (Dirichlet) by between 96.5% and 96.8% for $\omega = 5$ and between 96.7% and 96.9% for $\omega = 10$. The effect of the boundary condition may be interpreted in the following way. As a result of the transformation, the cloaked object and cloak together behave as if the void is small. In this sense, the cloaked inclusion represents a singular perturbation of the fundamental solution of the Helmholtz equation. In the case of a free void with Neumann conditions, the leading order term in the asymptotic expansion is the dipole term, which is of order ε^2 and decays like the first derivative of the fundamental solution. On the other hand, for a fixed void with Dirichlet conditions, the leading order term in the expansion is the monopole term which is of order ε and decays like the fundamental solution. Thus, the perturbation from the free void is smaller than the perturbation from the fixed void, leading to improved cloaking.

5.4 Cloaking in elastic plates

Now, we extend the concept of the square regularised cloak to the problem of wave propagation for Kirchhoff–Love plates. The governing equations change, as described below. Also, the new formulation requires a new set of boundary conditions, or transmission conditions on the interface. As emphasised at the beginning of the chapter, particular attention is devoted to the boundary conditions on the interior boundary of the regularised cloak. In particular, it will be shown that the rigid clamping of the interior boundary boundary makes the cloaking in flexural elastic waves impossible. The results presented in this section are based on the work reported in [45, 81].

In the absence of applied in-plane forces, the equation governing the time-harmonic out-of-plane displacement amplitude $w(\boldsymbol{X})$ of an isotropic homogeneous Kirchhoff–Love plate under pure bending is [96, 183]

$$\left(\nabla_{\boldsymbol{X}}^4 - \frac{Ph}{D^{(0)}}\omega^2\right)w(\boldsymbol{X}) = 0, \ \boldsymbol{X} \in \chi \subseteq \mathbb{R}^2, \tag{5.27}$$

where $D^{(0)}$, P, and h are the flexural rigidity, density, and thickness of the plate respectively; and ω is the radian frequency. Consider an invertible transformation $\mathcal{F} : \chi \mapsto \Omega$ and $\boldsymbol{x} = \mathcal{F}(\boldsymbol{X})$. By a double application of [141, Lemma 2.1] Equation (5.27) in new coordinates may be expressed as

$$\left(\nabla \cdot J^{-1}\boldsymbol{F}\boldsymbol{F}^T\nabla J \nabla \cdot J^{-1}\boldsymbol{F}\boldsymbol{F}^T\nabla - \frac{Ph}{JD^{(0)}}\omega^2\right)w(\boldsymbol{x}) = 0, \ \boldsymbol{x} \in \Omega, \tag{5.28}$$

where $\nabla = \nabla_{\boldsymbol{x}}$, $\boldsymbol{F} = \nabla_{\boldsymbol{X}}\boldsymbol{x}$ is the deformation gradient and $J = \det \boldsymbol{F}$ is the Jacobian. For ease of exposition in what follows it is convenient to work in Cartesian coordinates. The differentiation in the equations below is applied with respect to the vector variable $\boldsymbol{x}$ in the transformed domain. Using the Einstein summation convention the transformed equation (5.28) expressed in Cartesian coordinates is

$$JG_{ij}G_{kl}\,w_{,ijkl} + 2\left(JG_{ij}G_{kl}\right)_{,i}w_{,jkl}$$
$$+ \left[G_{ij}(JG_{kl})_{,ij} + 2G_{jk}\left(JG_{il,i}\right)_{,j} + G_{ij,i}(JG_{kl})_{,j} + JG_{ik,i}G_{jl,j}\right]w_{,kl}$$
$$+ \left[G_{ij,i}\left(JG_{kl,k}\right)_{,j} + G_{ij}\left(JG_{kl,k}\right)_{,ij}\right]w_{,l} - \frac{Ph}{JD^{(0)}}\omega^2 w = 0, \tag{5.29}$$

where the symmetric tensor

$$G_{ij} = J^{-1}F_{ip}F_{jp} \tag{5.30}$$

has been introduced and subscript commas followed by indices indicate differentiation with respect to spatial variables. It is emphasised that the governing equation for flexural vibrations in a thin elastic plate is not invariant under coordinate transformation as can be seen from (5.29).

Hence, there is a difference, compared to a similar procedure applied to the transformation of the Helmholtz equation, as used in models of invisibility cloaks in acoustics or electromagnetism (see, for example, [57, 98, 141, 158]). In particular, the transformed Helmholtz equation, with the appropriate choice of the gauge, can be interpreted as the governing equation for time-harmonic waves in an inhomogeneous anisotropic medium. There was no need for the introduction of any additional external fields in such models. In contrast, Equation (5.29) is not the standard form for an anisotropic inhomogeneous plate. Namely, as in [96], the equation of anisotropic inhomogeneous plate, of flexural rigidities $D_{ijkl}(\boldsymbol{x})$, has the form

$$D_{ijkl}w_{,ijkl} + 2D_{ijkl,i}w_{,jkl} + D_{ijkl,ij}w_{,kl} - h\rho\omega^2 w = 0. \qquad (5.31)$$

Whereas the first two terms in (5.31) have the same structure as in (5.29), there is a discrepancy in the structure of the remaining terms involving first and second-order derivatives of w in the above Equations (5.29) and (5.31). This discrepancy cannot be rectified by any choice of elastic constants or inertia in the transformed domain.

A physical interpretation can be given to the transformed equation (5.29), subject to the introduction of an appropriate pre-stress. Furthermore, this approach will be used to construct an invisibility cloak for flexural waves within the framework of anisotropic pre-stressed plate theory.

As in [48, 141, 143], the transformations considered here are assumed to be invertible. *Perfect* cloaks require that the transformation be singular on the inner boundary of the cloak, which leads to cloaks with singular material properties. However, a regularisation procedure can lead to construction of *near cloaks* as introduced by Kohn et al. [88]. The regularisation parameter ε can be taken as small as desired in order to achieve the required accuracy of the cloak.

5.4.1 Governing equations in the presence of in-plane forces

The following discussion provides the physical interpretation of Equation (5.29) in the transformed domain. In the presence of in-plane forces, the time-harmonic flexural deformation of a Kirchhoff–Love plate is governed by the following equation (see [95, 96, 183]):

$$M_{ij,ij} + N_{ij}w_{,ij} - S_i w_{,i} = -h\rho\omega^2 w, \qquad (5.32)$$

where w is the out-of-plane displacement. Here the plate is subjected to membrane forces (N_{ij}) and in-plane body forces (S_i). Additionally, the membrane and in-plane body forces are constrained to satisfy the equilibrium equation

$$N_{ij,j} + S_i = 0. \qquad (5.33)$$

Using the constitutive equation for a linearly elastic Kirchhoff–Love plate, $M_{ij} = -D_{ijkl}(\boldsymbol{x})w_{,kl}$, the equation of motion becomes

$$D_{ijkl}w_{,ijkl} + 2D_{ijkl,i}w_{,jkl} + (D_{ijkl,ij} - N_{kl})w_{,kl} + S_l w_{,l} = h\rho\omega^2 w, \qquad (5.34)$$

where D_{ijkl} are the flexural rigidities. Using Equation (5.34), the terms in the transformed equation (5.29) may be identified with physically meaningful quantities. In this manner, the transformed equation governing the flexural displacement of a Kirchhoff–Love plate under an arbitrary coordinate mapping may be interpreted as a generalised plate. It is emphasised that the generalised model remains within the framework of linear Kirchhoff–Love theory. In particular, the flexural rigidities of the transformed plate are immediately identified

$$D_{ijkl} = D^{(0)} J G_{ij} G_{kl}, \tag{5.35a}$$

as are the in-plane body forces

$$S_l = D^{(0)} \left[G_{jk} \left(J G_{il,i} \right)_{,j} \right]_{,k}, \tag{5.35b}$$

and the transformed density

$$\rho = \frac{P}{J}. \tag{5.35c}$$

It is clear that the flexural rigidities (5.35a) possess the expected major and minor symmetries. Hence, in general, there are six independent elastic parameters required to define a so-called "platonic" cloak. In addition to the above, the transformed equation (5.29) should also satisfy two additional constraints. Firstly, the coefficients of the third order terms in (5.29) must match the derivatives of the flexural rigidities, that is, $D_{ijkl,i} = D^{(0)} (J G_{ij} G_{kl})_{,i}$.

Secondly, the membrane forces N_{kl} must be chosen such that the second order terms in (5.29) match those in (5.34). Further, the membrane and in-plane body forces must also satisfy the equilibrium equation (5.33). The membrane forces appear only in terms involving the second order derivative and it is assumed that w is sufficiently smooth to allow the order of differentiation to be interchanged. These forces are obtained by integrating the stresses through the thickness of the plate

$$N_{kl} = \int_{-h/2}^{h/2} \sigma_{kl} \, dz,$$

whence $N_{kl} = N_{lk}$ is required for symmetric stress. The desired symmetry is obtained by taking N_{kl} in the form

$$N_{kl} = D^{(0)} \left[\left(J G_{kl} G_{ij,i} - J G_{jl} G_{ik,i} \right)_{,j} - G_{jk} \left(J G_{il,i} \right)_{,j} \right]. \tag{5.35d}$$

It is now straightforward to verify that the membrane and body forces satisfy the in-plane equilibrium equation (5.33).

It has thus been demonstrated that, under a general coordinate mapping, the equation governing time-harmonic flexural vibrations of a linear isotropic homogeneous Kirchhoff–Love plate transforms to an equation corresponding to a linear anisotropic inhomogeneous Kirchhoff–Love plate in the presence of in-plane loads. It is emphasised that these loads depend only on the coordinate mapping (via the deformation gradient) and are not functions of displacement nor time. In this sense, the membrane forces N_{kl} can be interpreted as a pre-stress together with appropriate body forces S_l to ensure equilibrium. This formalism represents a general framework in which transformation elastodynamics for Kirchhoff–Love plates can be investigated. The distinguishing feature of this interpretation is that, although a generalised plate model is introduced, the framework is entirely linear and all terms are identified with well understood physical quantities.

It is observed that a compressive pre-stress may occur, as demonstrated in Section 5.4.4, and, as a result, the question of buckling may arise in the practical implementation. The buckling load depends not only on the geometrical and material parameters of the structure, but also on the distribution of the pre-stress arising from the particular geometric transformation implemented. The maximum compressive stress occurs at the interior boundary of the cloak in neighbourhoods of any corner points and its magnitude is governed by the regularisation parameter ε used to create the near cloaks.

A further notable feature of the new framework, as above, is that it ensures that the stiffnesses have both major and minor symmetries, the stresses are symmetric and the transformed density is scalar. This is in contrast to the case of cloaks for vector three- and two-dimensional elasticity [143] where there is either a non-symmetric stress [28] or tensorial mass density and dependence of stress on velocity [119]. It is also appropriate to mention the use of pre-stress in problems of control of elastic waves for out-of-plane and in-plane wave motion in hyperelastic materials [142, 156, 157] and interface incremental problems of vector elasticity with finite pre-stress [20]. In the latter case, the emphasis is on the modelling and control of Bloch–Floquet waves in a surface elastic layer subjected to a finite pre-stress and localisation near defects, rather than re-routing of elastic waves around an obstacle.

5.4.2 Interface conditions

Field equations (5.28) or (5.34) are accompanied by transmission conditions on the interface boundary between transformed and untransformed domains. The interface conditions can be deduced by applying the principle of virtual displacement (see [183] among others) and extending the result to anisotropic media. The essential interface conditions are the continuity of transverse displacement and its normal derivative on the interface between two domains:

$$[\![w]\!] = 0, \quad \left[\!\!\left[\frac{\partial w}{\partial n} \right]\!\!\right] = 0, \quad \text{on } \partial\Omega, \qquad (5.36)$$

where $[\![.]\!]$ denotes the jump and $\mathbf{n}$ is the unit normal on the interface $\partial\Omega$. The natural interface conditions correspond to the continuity of the vertical forces and of the normal component of the bending moment across the domain

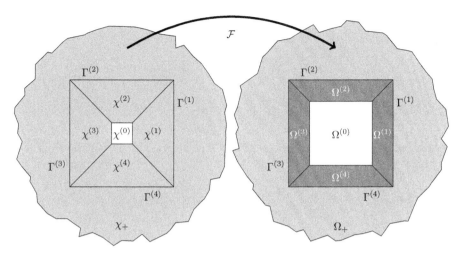

FIGURE 5.13: The map $\mathcal{F}$ transforms the undeformed region $\chi = \left(\cup_{i=1}^{4}\chi^{(i)}\right) \cup \chi_{+}$ to the deformed configuration $\Omega = \left(\cup_{i=1}^{4}\Omega^{(i)}\right) \cup \Omega_{+}$. The exterior of the domain remains unchanged by the transformation such that $\Omega_{+} = \chi_{+}$. The cloak is shaded in dark grey in the deformed configuration.

interface. They can be expressed in term of $w(\mathbf{x})$ in the form

$$\left[\!\!\left[\left(N_{ij}\,w_{,j} - (D_{ijkl}w_{,kl})_{,j} - \frac{\partial}{\partial s}(D_{ijkl}w_{,kl}s_{j})\right)n_{i}\right]\!\!\right] = 0, \tag{5.37}$$

$$\left[\!\!\left[\,D_{ijkl}w_{,kl}\,n_{j}n_{i}\right]\!\!\right] = 0, \quad \text{on } \partial\Omega,$$

where s is the counter-clockwise unit tangent on $\partial\Omega$.

5.4.3 Square cloak

Here, the square cloak concept, already discussed for the membrane problem in Section 5.2, is extended to the equations governing waves in Kirchhoff–Love plates. Geometrically, the coordinate transformation deforms a small square $\chi^{(0)} = \{\mathbf{X} : |X_1|/a < \varepsilon, |X_2|/a < \varepsilon\}$ together with the surrounding four trapezoids χ_i, into a larger square $\Omega^{(0)} = \{\mathbf{x} : |x_1| < a, |x_2| < a\}$ and four narrower trapezoids $\Omega^{(i)}$ as illustrated in Figure 5.13. The invisibility cloak is then formed from the four deformed trapezoids $\Omega_{-} = \cup_{i=1}^{4}\Omega^{(i)}$ and surrounds the cloaked region $\Omega^{(0)}$. The map transforms the original domain $\chi = \left(\cup_{i=0}^{4}\chi^{(i)}\right) \cup \chi_{+}$ to the deformed configuration $\Omega = \Omega_{+} \cup \Omega_{-}$ in such a way that leaves the exterior of the cloak unchanged, that is, $\mathcal{F}(\chi_{+}) = \chi_{+} = \Omega_{+}$ and $\mathcal{F}(\Gamma^{(i)}) = \Gamma^{(i)}$.

The mapping is defined in such a way that $\mathcal{F}(\mathbf{X}) = \mathcal{F}^{(i)}(\mathbf{X})$ for $\mathbf{X} \in \chi^{(i)}$ ($i = 1, \ldots, 4$) and is the identity on $\chi_{+} = \Omega_{+}$. In particular, the mapping, the deformation gradient and the Jacobian for the regions $\Omega^{(1)}$, $\Omega^{(3)}$ are

$$\mathcal{F}^{(1,3)} = \begin{pmatrix} \alpha_1 X_1 \pm \alpha_2 \\ \alpha_1 X_2 \pm \alpha_2 X_2/X_1 \end{pmatrix}, \quad \boldsymbol{F}^{(1,3)} = \begin{pmatrix} \alpha_1 & 0 \\ \dfrac{x_2 \alpha_1 \alpha_2}{x_1(\alpha_2 \mp x_1)} & \dfrac{x_1 \alpha_1}{x_1 \mp \alpha_2} \end{pmatrix},$$

$$J^{(1,3)} = \frac{x_1 \alpha_1^2}{x_1 \mp \alpha_2},$$

where $\alpha_1 = b/(a+b-\varepsilon)$, $\alpha_2 = (a+b)(a-\varepsilon)/(a+b-\varepsilon)$, and b is the thickness of the cloak. Similarly, for $\Omega^{(2)}$, $\Omega^{(4)}$,

$$\mathcal{F}^{(2,4)} = \begin{pmatrix} \alpha_1 X_1 \pm \alpha_2 X_1/X_2 \\ \alpha_1 X_2 \pm \alpha_2 \end{pmatrix}, \quad \boldsymbol{F}^{(2,4)} = \begin{pmatrix} \dfrac{x_2 \alpha_1}{x_2 \mp \alpha_2} & \dfrac{x_1 \alpha_1 \alpha_2}{x_2(\alpha_2 \mp x_2)} \\ 0 & \alpha_1 \end{pmatrix},$$

$$J^{(2,4)} = \frac{x_2 \alpha_1^2}{x_2 \mp \alpha_2}.$$

5.4.4 Material parameters and pre-stress for the cloak

The corresponding material parameters and forces are obtained from the general formalism (5.35). For the right-hand side of the cloak, the six independent components of flexural rigidity are

$$D^{(1)}_{1111} = \alpha_1^2 \left(1 - \frac{\alpha_2}{x_1}\right) D^{(0)},$$

$$D^{(1)}_{2222} = \frac{\alpha_1^2 \left(\alpha_2^2 x_2^2 + x_1^4\right)^2}{(x_1 - \alpha_2)^3 x_1^5} D^{(0)},$$

$$D^{(1)}_{2211} = \frac{\alpha_1^2 \left(\alpha_2^2 x_2^2 + x_1^4\right)}{(x_1 - \alpha_2) x_1^3} D^{(0)},$$

$$D^{(1)}_{1212} = \frac{\alpha_1^2 \alpha_2^2 x_2^2}{(x_1 - \alpha_2) x_1^3} D^{(0)},$$

$$D^{(1)}_{1112} = -\alpha_1^2 \alpha_2 \frac{x_2}{x_1^2} D^{(0)},$$

$$D^{(1)}_{2212} = -\frac{\alpha_1^2 \alpha_2 x_2 \left(\alpha_2^2 x_2^2 + x_1^4\right)}{(x_1 - \alpha_2)^2 x_1^4} D^{(0)}.$$

The remaining components can be deduced from the major and minor symmetries of $\boldsymbol{D}$. The membrane and body forces are

$$N^{(1)}_{11} = \frac{2\alpha_1^2 \alpha_2}{x_1^2(x_1 - \alpha_2)} D^{(0)}, \quad N^{(1)}_{12} = \frac{2\alpha_1^2 \alpha_2 x_2 (3x_1 - 2\alpha_2)}{(x_1 - \alpha_2)^2 x_1^3} D^{(0)},$$

$$N^{(1)}_{22} = -\frac{2\alpha_1^2 \alpha_2 \left(x_1^4 + 8\alpha_2 x_2^2 x_1 - 3\alpha_2^2 x_2^2\right)}{x_1^4 (x_1 - \alpha_2)^3} D^{(0)},$$

$$S_1^{(1)} = 0, \quad S_2^{(1)} = \frac{24\alpha_1^2 \alpha_2 x_2}{(x_1 - \alpha_2)^3 x_1^2} D^{(0)}.$$

Finally, the density is

$$\rho^{(1)} = \frac{P(x_1 - \alpha_2)}{\alpha_1^2 x_1}.$$

The corresponding physical quantities for the remaining three sides can be deduced through symmetry. The material on the interior of the cloak corresponds to an inhomogeneous anisotropic Kirchhoff–Love plate with inhomogeneous and anisotropic pre-stress. All material parameters and forces are finite for $\varepsilon > 0$.

5.4.4.1 Principal directions of orthotropy

At each point of the plate, one can introduce a system of coordinates, which coincides with the principal directions of orthotropy of the plate. To this end, a local angle of rotation θ is introduced. The rotation is considered local in the sense that the angle of rotation $\theta = \theta(x)$ is a function of position. For convenience, the following reduced form for the six independent components of flexural rigidities are introduced

$$D_{11}^{(i)} = D_{1111}^{(i)}, \quad D_{12}^{(i)} = D_{1122}^{(i)}, \quad D_{16}^{(i)} = D_{1112}^{(i)},$$
$$D_{22}^{(i)} = D_{2222}^{(i)}, \quad D_{26}^{(i)} = D_{2221}^{(i)}, \quad D_{66}^{(i)} = D_{1212}^{(i)}.$$

The reduced flexural rigidities $D_{16}^{(i)}$ and $D_{26}^{(i)}$, sometimes called *auxiliary rigidities*, vanish in the principal directions of the plate (see [96, 183] among others). Consider a *local* rotation of the coordinate system $\mathcal{G} : x \mapsto \tilde{x}$. In this new coordinate system, the auxiliary rigidities may be expressed as

$$\begin{aligned}
\tilde{D}_{16}^{(i)} &= \frac{1}{2}\left[D_{22}^{(i)} \sin^2\theta - D_{11}^{(i)} \cos^2\theta + \left(D_{12}^{(i)} + 2D_{66}^{(i)}\right)\cos 2\theta\right]\sin 2\theta \\
&+ D_{16}^{(i)} \cos^2\theta\left(\cos^2\theta - 3\sin^2\theta\right) + D_{26}^{(i)} \sin^2\theta\left(3\cos^2\theta - \sin^2\theta\right), \\
\tilde{D}_{26}^{(i)} &= \frac{1}{2}\left[D_{22}^{(i)} \cos^2\theta - D_{11}^{(i)} \sin^2\theta - \left(D_{12}^{(i)} + 2D_{66}^{(i)}\right)\cos 2\theta\right]\sin 2\theta \\
&+ D_{16}^{(i)} \sin^2\theta\left(3\cos^2\theta - \sin^2\theta\right) + D_{26}^{(i)} \cos^2\theta\left(\cos^2\theta - 3\sin^2\theta\right),
\end{aligned} \quad (5.38)$$

where the dependence of θ and $D_{jk}^{(i)}$ on x has been omitted but is understood. The appropriate local angles of rotation $\theta(x)$, which satisfy the above system of transcendental equations yields the local principal directions of rigidity. These principal directions for the square push-out transformation applied to the equations governing the flexural displacement of a Kirchhoff–Love plate are shown in Figure 5.14.

If $D_{16}^{(i)} = D_{26}^{(i)} = 0$, then the system of coordinates is already aligned with principal directions of orthotropy ($\theta = 0$). If additional non-zero rotation is needed ($\theta \neq 0$), then Equations (5.38) can be written in the form:

$$\tilde{D}_{16}^{(i)} + \tilde{D}_{26}^{(i)} = \frac{1}{2}\left(D_{22}^{(i)} - D_{11}^{(i)}\right)\sin 2\theta + \left(D_{16}^{(i)} + D_{26}^{(i)}\right)\cos 2\theta,$$

$$\tilde{D}_{26}^{(i)} - \tilde{D}_{16}^{(i)} = \frac{1}{2}\left(D_{11}^{(i)} + D_{22}^{(i)} - 2D_{12}^{(i)} - 4D_{66}^{(i)}\right)\sin 2\theta \cos 2\theta \qquad (5.39)$$
$$+ \left(D_{26}^{(i)} - D_{16}^{(i)}\right)\left(\cos^2 2\theta - \sin^2 2\theta\right),$$

which have to be set equal to zero in order to find the principal direction angle $\theta \in (0, \pi/2]$. Substituting the solution of the first equation in (5.39)

$$\sin 2\theta = 2\frac{D_{16}^{(i)} + D_{26}^{(i)}}{D_{11}^{(i)} - D_{22}^{(i)}}\cos 2\theta, \qquad (5.40)$$

into the second one, leads to

$$\cos 4\theta = -\frac{\mathcal{D}}{\mathcal{D} + 2\left(D_{11}^{(i)} - D_{22}^{(i)}\right)\left(D_{26}^{(i)} - D_{16}^{(i)}\right)}, \qquad (5.41)$$

where $\mathcal{D} = (D_{11}^{(i)} + D_{22}^{(i)} - 2D_{12}^{(i)} - 4D_{66}^{(i)})(D_{16}^{(i)} + D_{26}^{(i)})$. It is also observed that, for this angle of rotation, the rigidity component $\tilde{D}_{66}^{(i)}$ vanishes.

This local orthotropy gives further important physical meaning to the material parameters of the cloak. In particular, the sign of some flexural rigidities changes across the symmetry lines in the global coordinate system. This change in sign can now be interpreted as simply a change in orientation of the principal axes of the material.

It is interesting to compare the principal directions in the plate and in the membrane cloaking problem, considered earlier in Section 5.2, where the same geometric transformation is used. The principal directions of orthotropy for a membrane cloak are characterised by the eigenvectors of $\boldsymbol{G}$ (cf. Equation (5.30)). If we identify the local angle between the eigenvectors of $\boldsymbol{G}$ and the standard Cartesian basis vectors with θ, it is straightforward to verify that such an angle satisfies Equations (5.38), and thus, the principal directions of orthotropy for the membrane cloaking problem are the same as those for the Kirchhoff–Love cloaking problem.

5.4.5 Implementation of the cloak for the flexural plate

In order to simulate a cloak around a scatterer in an infinite Kirchhoff–Love plate, perfectly matched layers are used in the vicinity of the exterior boundary of the computational domain. The implementation has been developed for the push-out cloaking transformation in the finite element software COMSOL Multiphysics®. A steel plate of thickness 10^{-3} m is considered and the following parameter values are chosen: $D^{(0)} = 19.23$ N · m, P = 7800 kg · m^{-3}, $h = 10^{-3}$ m, $a = b = 0.5$ m, $\varepsilon = 10^{-3}$ and $\omega = 314$ rad · s^{-1}. It is emphasised

FIGURE 5.14: The principal directions of the flexural rigidity tensor for the square cloak in a Kirchhoff–Love plate. Results are presented for parameter values: $a = 0.5$ m, $b = 0.5$ m and $\varepsilon = 10^{-3}$.

that these parameters are used to demonstrate broadband cloaking, which persists over a wide range of frequencies, geometrical and material parameter values, within the usual constraints of the Kirchhoff–Love plate model. Figure 5.15a shows the flexural displacement generated by a cylindrical source in a Kirchhoff–Love plate with a square void. Figure 5.15b shows the corresponding field when the void is surrounded by a square cloak, constructed as described in Section 5.4.3. Figure 5.15c shows the fields for the plots in Figures 5.15a and 5.15b together with Green's function for an infinite Kirchhoff–Love plate along the line passing through the source and centre of the cloak. Here, the Green's function represents the unperturbed field in the absence of the void and cloak. It is observed that, outside the cloaked region, the cloaked field and unperturbed field are almost coincident. Figure 5.15 clearly illustrates the effectiveness of this platonic cloak.

5.4.5.1 Green's functions and comparison with cloaking for the Helmholtz operator

It follows from the transformed equation (5.29) that, in general, inhomogeneous anisotropic cloaks for Kirchhoff–Love plates cannot be decomposed into Helmholtz and modified-Helmholtz parts, as in the case of homogenous and isotropic Kirchhoff–Love plates. Physically this means that, within the cloak, propagating and evanescent modes couple. However, if the ambient plate on the exterior of the cloak is homogeneous one would expect propagating modes, corresponding to solutions of the Helmholtz equation, to dominate outside the immediate neighbourhood of the cloak. The corresponding problem for the membrane (Helmholtz operator) and a square cloak was considered in Section 5.2. As in (5.1), in Cartesian coordinates the transformed Helmholtz

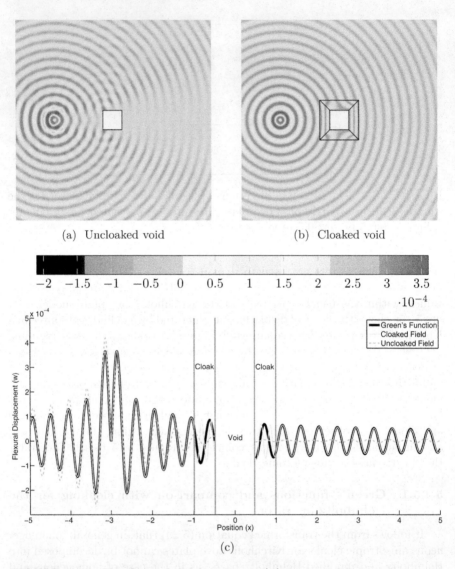

FIGURE 5.15: The flexural displacement $w(\boldsymbol{x})$ generated by a point source in the presence of an uncloaked (a) and cloaked (b) void. Figure (c) shows the flexural displacement for cases (a) and (b) together with the Green's function for a Kirchhoff–Love plate along the line passing through the source and centre of the cloak. The radian frequency is $\omega = 314 \text{ rad} \cdot \text{s}^{-1}$.

equation, which governs the out-of-plane displacement of a membrane, may be expressed as

$$G_{ij}u_{,ij} + G_{ij,i}u_{,j} + \frac{P_m\omega^2}{J\mu^{(0)}}u = 0,$$

where $\mu^{(0)}$ is the stiffness of the untransformed membrane and P_m the density. It is immediately apparent that, in general, solutions of the transformed Helmholtz equation will not satisfy the transformed equation for a plate (5.29). Nevertheless, if the exterior medium is homogeneous, one would expect solutions of the "corresponding" membrane problem to dominate in the plate solution at infinity. The Green's function $g_P(\boldsymbol{x})$ for Equation (5.27), which governs the flexural displacement of a Kirchhoff–Love plate, is [61]

$$g_P(\boldsymbol{x}) = i\frac{H_0^{(1)}(i\beta|\boldsymbol{x}|) - H_0^{(1)}(\beta|\boldsymbol{x}|)}{8\beta^2 D^{(0)}}, \tag{5.42}$$

where $\beta^4 = \omega^2 Ph/D^{(0)}$. The Green's function for the two-dimensional Helmholtz equation $g_H(\boldsymbol{x})$, which governs the out-of-plane displacement of a membrane, is

$$g_H(\boldsymbol{x}) = -i\frac{H_0^{(1)}(k|\boldsymbol{x}|)}{4\mu^{(0)}}, \tag{5.43}$$

where $k^2 = \omega^2 P_m/\mu^{(0)}$. For large arguments, the Hankel function $H_0^{(1)}$ has the following asymptotic representation [148]

$$H_0^{(1)}(z) \sim \sqrt{\frac{2}{\pi z}}e^{i(z-\frac{\pi}{4})},$$

as $z \to \infty$ in $-\pi + \delta \leq \arg(z) \leq 2\pi - \delta$, where $0 < \delta \ll 1$. Using the above representation yields the following expressions for the Green's functions at infinity,

$$g_P(\boldsymbol{x}) \sim -\frac{i}{8\beta^2 D^{(0)}}\sqrt{\frac{2}{\pi\beta|\boldsymbol{x}|}}e^{i(\beta|\boldsymbol{x}|-\frac{\pi}{4})},$$

and

$$g_H(\boldsymbol{x}) \sim -\frac{i}{4\mu^{(0)}}\sqrt{\frac{2}{\pi k|\boldsymbol{x}|}}e^{i(k|\boldsymbol{x}|-\frac{\pi}{4})},$$

whence is clear that $\beta = k$ is required for the two fields to share the same phase in the far field and choosing $2\beta^2 D^{(0)} = \mu^{(0)}$ gives equal amplitudes. Thus, for the cloaking problem for the plate and Helmholtz equation to have the same solution in the far field, the material parameters should be chosen such that $P_m = 2Ph$ and $\mu^{(0)} = 2\omega\sqrt{PhD^{(0)}}$. It is in this sense that the cloaking problem for the membrane is said to "correspond" to the cloaking problem for the plate.

Figure 5.16 shows the solution of the cloaking problem for the corresponding membrane problem (see [48]), and the difference between the fields for the

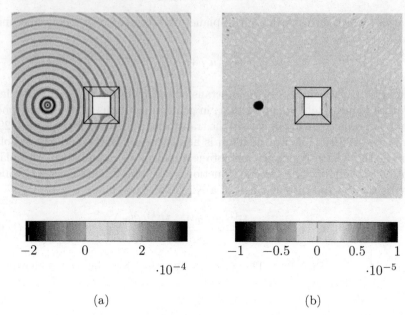

FIGURE 5.16: (a) The solution for the corresponding membrane cloaking problem, (b) the difference between the solution for the cloaking problem for plates and the corresponding membrane problem.

plate and membrane. In Figure 5.17a, the two solutions for the two cloaking problems (plate and membrane) are plotted along the line passing through the source and centre of the cloak. Figure 5.17b shows the difference between the two solutions along the same line. The reader's attention is drawn to the different scales in Figures 5.15, 5.16 and 5.17. It is also emphasised that the Green's function for the membrane problem (Helmholtz operator) (5.43) is singular at the origin whereas the fundamental solution for the plate equation (5.42) is regular, hence the large discrepancy in the vicinity of the source in Figures 5.15, 5.16 and 5.17 is justified. It is observed that, away from the source and outside the cloak, the difference between the solutions for the plate problem and corresponding problem for the membrane is small. However, this difference is significant on the interior of the cloak, particularly close to the inner boundary of the cloak.

5.4.5.2 Quality of cloaking

The scattering measure introduced in Section 5.2.1.5 will be used as a tool to quantify the effectiveness of the cloak:

$$\mathcal{E}(u_1, u_0, \mathcal{R}) = \left(\int_{\mathcal{R}} |u_1(\boldsymbol{x}) - u_0(\boldsymbol{x})|^2 \, \mathrm{d}\boldsymbol{x} \right) \left(\int_{\mathcal{R}} |u_0(\boldsymbol{x})|^2 \, \mathrm{d}\boldsymbol{x} \right)^{-1}, \quad (5.44)$$

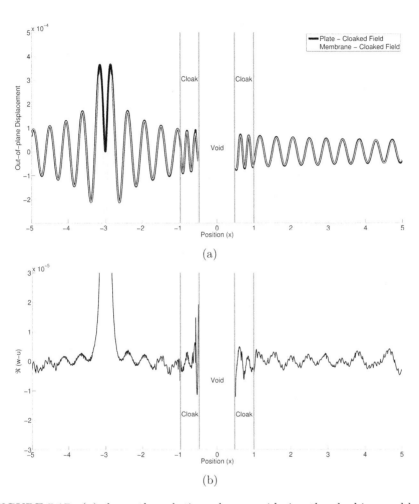

FIGURE 5.17: (a) shows the solution when considering the cloaking problem for plates and the corresponding membrane problem. (b) shows the difference between the solutions when considering the cloaking problem for plates and the corresponding membrane problem. In (a) and (b), the fields are plotted along the line passing through the source and centre of the cloak. The reader's attention is drawn to the different scales in (a) and (b).

where $\mathcal{R} \subset \mathbb{R}^2$ is also introduced in Section 5.2.1.5. In the numerical simulations, shown in Figures 5.15a and 5.15b, $\mathcal{R} = \mathcal{R}_1$ is used. It is emphasised that this choice of $\mathcal{R}$ yields a strict measure of the efficacy of the cloak; with not only forward and backward scattering effects accounted for, but also significant near field effects in the immediate neighbourhood of the cloak. The scattering measure for the uncloaked field shown in Figure 5.15a is $\mathcal{E}(w_u, u_0, \mathcal{R}) = 0.109$, whereas the scattering measure for the cloaked field shown in Figure 5.15b is $\mathcal{E}(w_c, u_0, \mathcal{R}) = 5.05 \times 10^{-4}$. The difference between the two scattering measures serves to emphasise the efficiency of the cloak.

5.4.5.3   Measuring cloaking quality for interference patterns

Interferometry is known as a possible method through which the quality of cloaks may be assessed [48]. In the earlier Section 5.2 the efficacy of a square cloak for Helmholtz waves was analysed using a Young's double-slit interferometer. It was also demonstrated that there is virtually no perturbation to the interference fringes when an aperture is covered with a cloaked object, whereas the uncloaked object severely distorts the interference pattern.

Here, the effectiveness of the cloak is examined by a similar method. Figure 5.18 shows the results of an interferometry simulation where the stability of the interference pattern generated by two coherent cylindrical sources is examined. Figure 5.18a shows the interference pattern generated by two coherent cylindrical sources in an infinite homogeneous plate. Figure 5.18b shows the perturbation to the interference pattern when a void is introduced, while Figure 5.18c shows the interference pattern when the void is coated with a cloak. The interference patterns seen on the observation screen (dashed lines in Figures 5.18a–5.18c) for all three cases are shown in Figure 5.18d. It is observed that the interference patterns for the homogeneous plate and cloaked void are virtually identical, whereas the presence of the uncloaked void significantly perturbs the interference pattern. It is emphasised that this is not a low-frequency response and the dimensions of cloak and inclusion are not small compared to the wavelength.

It is remarked that, in both cases, for the membrane waves and for flexural waves in Kirchhoff–Love plates, the interference pattern is very sensitive to perturbations and hence makes a good tool in the measurement of cloaking quality.

An experimental implementation of the structured cloak in the square flexural lattice is described in [125]. It has been demonstrated that the regularised cloak suppresses the scattered field within a wide frequency range in a plate containing a square hole. The paper [125] presents the results in the cymatics environment, which provides an elegant visualisation of the wave scattering in the time-harmonic regime.

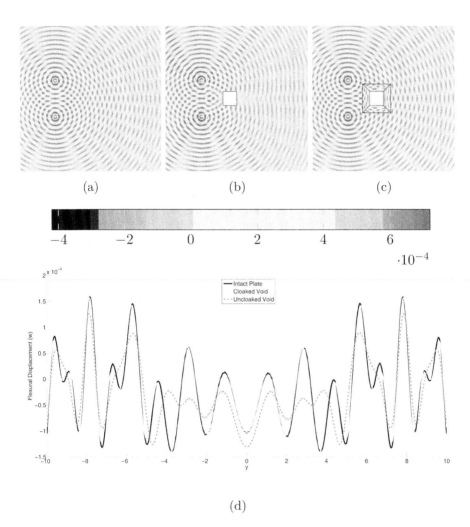

FIGURE 5.18: The flexural displacement generated by two coherent point sources in (a) an intact plate, (b) a plate with a square void, and (c) a plate with a cloaked square void. (d) shows the interference fringes along an observation screen, indicated by dashed lines in (a)–(c). The following parameter values were used: $D^{(0)} = 19.23$ Nm, P = 7800 kg/m^3, $h = 10^{-3}$ m, $a = b = 1.0$ m, $\varepsilon = 10^{-3}$ and $\omega = 157$ rad/s.

5.5 Singular perturbation analysis of an approximate cloak

This section is based on the results of the paper [81]. The matter of boundary conditions on the interior boundary of a cloaking region is rarely discussed and numerical simulations for cloaking are commonly presented without additional comments regarding these boundary conditions. The reason is that for a singular map, which "stretches" a hole of zero radius into a finite disk, in the unperturbed configuration there is no boundary (see, for example, [63–65,67]). Nevertheless, in the subsequent numerical computations the singularity of the material constants in the cloaking region is replaced by regularised finite values. Namely, the boundary conditions are required for the numerical computations and usually natural boundary conditions that follow from the variational formulation are chosen, i.e. these are Neumann boundary conditions on the interior contour of the cloaking region [142, 155, 156].

One can ask a naive question related to an illusion rather than cloaking. For example, if one has a large carrot cake, would it be possible to make it look like a small fairy cake instead. In turn, could one make a large void in a solid look smaller? Purely naïvely, the latter can be addressed through a geometrical transformation in polar coordinates:

$$r = \alpha_1 + \alpha_2 R, \quad \theta = \Theta,$$

where

$$\alpha_1 = \frac{R_2(R_1 - a)}{R_2 - a}, \quad \alpha_2 = \frac{R_2 - R_1}{R_2 - a},$$

and

$$a \leq R \leq R_2, \quad R_1 \leq r \leq R_2,$$

with a being a small positive number and (R, Θ) being the coordinates before the transformation and (r, θ) after the transformation, and R_1, R_2 being positive constants, such that $R_1 < R_2$. Such a transformation, within the ring $R_1 < r < R_2$, would correspond to a radial non-uniform stretch, and assuming that outside the disk $r > R_2$ there is no deformation (i.e. $r = R$), we obtain a coating that would lead to an illusion regarding the size of a defect in a solid. Push-out transformations of such a type have been widely used, and have been applied to problems of cloaking, for example, in papers [75, 98].

Formally, if u represents an undistorted field in the exterior of a small void, of radius a and we assume the series representation

$$u(R, \Theta) = \sum_{n=-\infty}^{\infty} V_n(R) W_n(\Theta),$$

with V_n and W_n being basis functions, then the distortion, represented by the fields u_1 and u_2 can be described through a "shifted" series representation, as follows

$$u_1(r,\theta) = \sum_{n=-\infty}^{\infty} V_n\left(\frac{r-\alpha_1}{\alpha_2}\right) W_n(\theta), \text{ when } R_1 < r < R_2$$

and

$$u_2(R,\Theta) = \sum_{n=-\infty}^{\infty} V_n(R) W_n(\Theta), \text{ when } r > R_2.$$

Such a transformation delivers an illusion, which makes a finite circular void of radius R_1, look like a small void of radius a.

The above argument may appear to be simplistic, but we are going to show that it works exactly as described, for a class of problems governed by the Helmholtz operator and by the equations of vibrating Kirchhoff–Love plates. The basis functions V_n and W_n are written in the closed form, as further shown in the text of this section.

We consider the problem in the framework of singular perturbations and, instead of regularising the singular values of material parameters after the cloaking transformation, we begin by introducing a small hole and apply the cloaking transformation to a region containing such a small hole with appropriate boundary conditions already chosen. This is consistent with the analysis presented in [48, 88].

We show that the cloaking problem is closely related to that for an infinite body containing a small hole of radius a where we set a boundary condition of either the Dirichlet (clamped boundaries) or Neumann (free-edge boundaries) type. This is done for both membrane waves governed by the Helmholtz equation and flexural waves that occur in Kirchhoff–Love plates. Analytical solutions are presented for the cloaking problems and it is shown that the degree of cloaking is highly dependent on the boundary condition on the interior contour of the cloaking region. In a plate, in the presence of an incident plane wave along the x-axis, the scattered flexural field, u_s, outside the cloaking region, when $a \to 0$ with a clamped interior boundary, has the asymptotic representation at a sufficiently large distance R from the centre of the scatterer

$$u_s \sim -\sqrt{\frac{2}{\pi \beta R}} \exp\left(i\left(\beta R - \frac{\pi}{4}\right)\right), \text{ as } \beta R \to \infty, \tag{5.45}$$

where $\beta^4 = \rho h \omega^2 / D_0$ with radian frequency ω, plate flexural rigidity D_0, plate thickness h and density ρ. This immediately suggests the absence of any cloaking action as the above asymptotic representation corresponds to a finite point force initiated by a rigid pin at the origin (see [61, 111]). On the contrary, the free-edge interior boundary in the cloaking layer for a flexural plate produces high-quality cloaking action, and the corresponding asymptotic representation of the scattered field u_s at sufficiently large βR becomes

$$u_s \sim \sqrt{\frac{2}{\pi \beta R}} \exp\left(i\left(\beta R - \frac{\pi}{4}\right)\right) \frac{\pi i \nu}{4(1-\nu)}(\beta a)^2 = O(\beta a)^2, \qquad (5.46)$$

as $\beta a \to 0$ and $\beta R \to \infty$.

This confirms that the regularisation algorithm that refers to a small free-edge hole produces a scattered field proportional to the area of this small hole, and it tends to zero as $\beta a \to 0$.

For the "cloaking type" map considered here, the original hole does not have to be small, and we will refer to a "cloaking illusion", which results in an object being mimicked by an obstacle of a different size.

In Section 5.5.1 we introduce the notion of a cloaking transformation. Section 5.5.3 addresses the singular perturbation approach for an elastic membrane. In particular, in Section 5.5.3.1 we include an analytical solution, together with asymptotic estimates, for the model problem of scattering of membrane waves from a small circular scatterer. Section 5.5.3.2 shows the relationship between the model problem for a body with a small scatterer and the full cloaked problem. Section 5.5.4 presents the analytical solution and the asymptotic analysis of scattering of flexural waves in a Kirchhoff–Love plate for the biharmonic cloaking problem.

Based on the model of a near-cloak, for flexural waves in a Kirchhoff–Love plate it is shown that the object surrounded by the cloak, appears as an infinitesimally small scatterer, with appropriate boundary conditions on its contour. The perception of a scattered field, that vanishes when the diameter of such a scatterer tends to zero, does not apply to the case of flexural waves scattered by a clamped small inclusion.

Even when a singular transformation is used to design an "exact" cloak, in the numerical implementation, or in a physical experiment, it is always regularised. Of course, such regularisation is linked to setting boundary conditions on the interior contour of the cloaking region.

5.5.1 Push-out transformation

A non-conformal transformation is introduced to define the cloaking coating. We use the notations $\boldsymbol{X} = (R, \Theta)^T$ and $\boldsymbol{x} = (r, \theta)^T$ for coordinates before the transformation and after the transformation, respectively. Here (R, Θ) are polar coordinates. With reference to [75, 98, 141, 158, 173], we use the radial invertible "push-out" map $\mathcal{F}: \boldsymbol{X} \to \boldsymbol{x}$ defining the new stretched coordinates (r, θ) as follows. When $|\boldsymbol{X}| < R_2$, the transformation $\boldsymbol{x} = \mathcal{F}(\boldsymbol{X})$ is given by

$$r = R_1 + \frac{(R_2 - R_1)}{R_2}R, \quad \theta = \Theta, \qquad (5.47)$$

so that $R_1 < r < R_2$ when $0 < R < R_2$, with R_1 and R_2 being the interior and exterior radii of the cloaking ring, respectively.

In the exterior of the cloaking region, when $|\boldsymbol{X}| > R_2$, the transformation map is defined as the identity, so that $\boldsymbol{x} = \boldsymbol{X}$.

In the cloaking region, the Jacobi matrix $\mathbf{F} = D\mathcal{F}/D\boldsymbol{X}$ has the form

$$\mathbf{F} = \frac{R_2 - R_1}{R_2}\mathbf{e}_r \otimes \mathbf{e}_r + \frac{R_2 - R_1}{R_2}\frac{r}{r - R_1}\mathbf{e}_\theta \otimes \mathbf{e}_\theta, \qquad (5.48)$$

where $\mathbf{e}_r = \mathbf{e}_R$, $\mathbf{e}_\theta = \mathbf{e}_\Theta$ is the cylindrical orthonormal basis and $\otimes$ stands for the dyadic product.

After the above transformation, the classical Laplace's operator will be modified accordingly. For a given real α, which is identified with R_1 for transformation (5.47), a new differential operator $\hat{\nabla}^2_\alpha$ is defined as

$$\hat{\nabla}^2_\alpha = \frac{1}{r-\alpha}\frac{\partial}{\partial r}\left[(r-\alpha)\frac{\partial}{\partial r}\right] + \frac{1}{(r-\alpha)^2}\frac{\partial^2}{\partial \theta^2}. \qquad (5.49)$$

As in [27], we refer to the operators $\hat{\nabla}^2_\alpha$, $\hat{\nabla}^2_\alpha + \beta^2$, and $\hat{\nabla}^2_\alpha - \beta^2$ as the "shifted Laplace", "shifted Helmholtz", and "shifted modified Helmholtz" operators, respectively.

5.5.2 Physical interpretation of transformation cloaking for a membrane

With the reference to the work of Norris [141], we give an outline of the transformed equation. Firstly, the governing equation for a time-harmonic out-of-plane displacement u of an elastic membrane has the form

$$\left(\nabla_{\boldsymbol{X}} \cdot \mu \nabla_{\boldsymbol{X}} + \rho\omega^2\right)u(\boldsymbol{X}) = 0, \quad \boldsymbol{X} \in \mathbb{R}^2, \qquad (5.50)$$

where μ, ρ, and ω represent the stiffness matrix, the mass density and the radian frequency respectively.

On application of the transformation $\boldsymbol{x} = \mathcal{F}(\boldsymbol{X})$ within the cloaking region, the transformed equation becomes

$$\left(\nabla \cdot \mu C(\boldsymbol{x})\nabla + \frac{\rho\omega^2}{J(\boldsymbol{x})}\right)u(\boldsymbol{x}) = 0, \qquad (5.51)$$

where

$$C = \frac{\mathbf{F}\mathbf{F}^{\mathrm{T}}}{J}, \quad \mathbf{F} = \nabla_{\boldsymbol{X}}\boldsymbol{x}, \quad J = \det \mathbf{F}. \qquad (5.52)$$

We note that Equation (5.51), similar to (5.50), describes a vibrating membrane, but with different elastic stiffness, which is inhomogeneous and orthotropic, and a non-uniform distribution of mass across the transformed region.

In contrast, for the model of a flexural plate, it was shown in [27, 45] that an additional pre-stress and body force are required to provide a full physical interpretation of the equations of motion in the transformed region, which again corresponds to an inhomogeneous and orthotropic material. The fourth order governing equation for the plate is to be discussed in Section 5.5.4. Special attention will be given to boundary conditions on the interior of the cloaking region.

Prior to the geometrical transformation, there was no need to prescribe any boundary conditions at the origin. On the other hand, after the transformation, one has an interior boundary of a finite size, and boundary conditions are required. In particular, this issue always occurs if numerical simulations are carried out for an invisibility cloak using the finite element method. On many occasions, such boundary conditions are chosen as natural boundary conditions and this means conditions of the Neumann type, which correspond to a free boundary at the interior of the cloaking layer.

The question arises as to whether or not imposing a Dirichlet condition on the interior boundary of the cloaking layer would make a difference in a numerical simulation or an experimental implementation of the cloak. The answer to this question is affirmative, and it is linked to the analysis of a class of singularly perturbed problems introduced below.

5.5.3 Singular perturbation problem in a membrane

It will be shown that a regularised cloak in a membrane mimics a small circular scatterer and furthermore, if the radius of such a scatterer tends to zero the perturbation of the incident field due to an interaction with the obstacle also reduces to zero. However, the asymptotic behaviour of the physical field around the scatterer, as the radius $a \to 0$, depends on the type of boundary conditions on the scatterer.

5.5.3.1 Model problem: scattering of a plane wave by a circular obstacle in a membrane

Consider an unbounded homogeneous isotropic elastic solid with shear modulus μ and density ρ containing a circular void of radius a. Further, incident time-harmonic elastic out-of-plane shear wave of frequency ω results in a total scattered time-harmonic field of amplitude $u(\mathbf{X})$ which obeys the Helmholtz equation outside the scatterer:

$$\left(\nabla^2 + k^2\right) u(\mathbf{X}) = 0, \tag{5.53}$$

where $k = \omega\sqrt{\rho/\mu}$.

We choose polar coordinates (R, Θ) such that the direction of the incident plane waves is along the $\Theta = 0$ line. The total displacement field, obeying the radiation condition at infinity, may be written as

$$u(R, \Theta) = \sum_{n=-\infty}^{\infty} \left[i^n J_n(kR) - i^n p_n H_n^{(1)}(kR)\right] e^{in\Theta}, \tag{5.54}$$

where J_n is the Bessel function of the first kind and $H_n^{(1)}$ is the first Hankel function. Use has also been made of the multipole expansion of the plane wave

$$e^{ikR\cos\Theta} = \sum_{n=-\infty}^{\infty} i^n J_n(kR) e^{in\Theta}. \tag{5.55}$$

The coefficients p_n are determined by the boundary condition on the circular scatterer. They are given by:

$$p_n = \begin{cases} \dfrac{J_n(ka)}{H_n^{(1)}(ka)} & : \text{if} \quad u = 0 \quad \text{on} \quad R = a, \\[2ex] \dfrac{J_n'(ka)}{H_n^{(1)\prime}(ka)} & : \text{if} \quad \dfrac{\partial u}{\partial R} = 0 \quad \text{on} \quad R = a. \end{cases} \tag{5.56}$$

Thus the field may be written as

$$u(R, \Theta) = e^{ikR\cos\Theta} - p_0 H_0^{(1)}(kR) - \sum_{n=1}^{\infty} i^n H_n^{(1)}(kR)(p_n e^{in\Theta} + p_{-n} e^{-in\Theta}) \tag{5.57}$$

For the Dirichlet boundary condition $u = 0$ on $R = a$, this may be rewritten as

$$u(R, \Theta) = e^{ikR\cos\Theta} - \frac{J_0(ka)}{H_0^{(1)}(ka)} H_0^{(1)}(kR) - \sum_{n=1}^{\infty} 2i^n \frac{J_n(ka)}{H_n^{(1)}(ka)} H_n^{(1)}(kR) \cos(n\Theta) \tag{5.58}$$

Correspondingly, for the case of the Neumann boundary condition $\partial u/\partial R = 0$ on $R = a$, we have

$$u(R, \Theta) = e^{ikR\cos\Theta} - \frac{J_0'(ka)}{H_0^{(1)\prime}(ka)} H_0^{(1)}(kR) - \sum_{n=1}^{\infty} 2i^n \frac{J_n'(ka)}{H_n^{(1)\prime}(ka)} H_n^{(1)}(kR) \cos(n\Theta) \tag{5.59}$$

In the asymptotic limit, when $ka \ll 1$, the coefficient of $H_0^{(1)}(kR)$ in the case of Dirichlet boundary conditions reduces to

$$-\frac{J_0(ka)}{H_0^{(1)}(ka)} \sim \frac{\pi i}{2\log(ka)} \quad \text{as } ka \to 0, \tag{5.60}$$

with the higher order coefficients tending to zero more quickly. In this case the scattered field is dominated by only the monopole term for $ka \to 0$.

For the Neumann boundary conditions in the asymptotic limit, as $ka \to 0$, the monopole and dipole coefficients are

$$-\frac{J_0'(ka)}{H_0^{(1)\prime}(ka)} \sim -\frac{\pi i}{4}(ka)^2, \tag{5.61}$$

and
$$-\frac{iJ_1'(ka)}{H_1^{(1)'}(ka)} \sim -\frac{\pi}{4}(ka)^2, \tag{5.62}$$

and they have leading order terms which are $O((ka)^2)$.

We note that in both above cases of boundary conditions, the coefficients tend to zero as $ka \to 0$. Furthermore, the coefficients of the higher-order multipole terms also vanish as $ka \to 0$. With that said, the monopole coefficient in (5.60), corresponding to the Dirichlet problem, is of order $O(|\log(ka)|^{-1})$, i.e. decays logarithmically slowly. Thus for a given size of small scatterer, the scattered field is reduced when Neumann boundary conditions are applied at the edge of the scatterer in comparison with the scattered field when Dirichlet boundary conditions are applied.

5.5.3.2 Boundary conditions and the cloaking problem in a membrane

We note that the components of the stiffness matrix $\boldsymbol{C}$ in (5.51) after the "cloaking transformation" (5.47) are singular at the interior boundary of the cloaking layer. In numerical computations, a regularisation is always introduced at the interior boundary to "eliminate" the singularity.

Alternatively, one can follow the approach advocated in [48] when a "near-cloak" is defined through the geometrical transformation of a plane with a small hole of radius a into a plane containing a ring, with the unperturbed exterior radius R_2 and an interior radius R_1. Outside the ring, the transformation is equal to the identity. Within the cloaking region, such a transformation is defined by

$$r = \alpha_1 + \alpha_2 R, \quad \theta = \Theta, \tag{5.63}$$

where
$$\alpha_1 = \frac{R_2(R_1 - a)}{R_2 - a}, \quad \alpha_2 = \frac{R_2 - R_1}{R_2 - a}. \tag{5.64}$$

and
$$a \leq R \leq R_2, \quad R_1 \leq r \leq R_2. \tag{5.65}$$

Outside the ring, the governing equation is (5.50), whereas inside the ring the equation has the form (5.51). For simplicity, assume that the material outside the ring is isotropic and homogeneous, i.e. $\mu = \text{const}$. Then, Equation (5.51) can also be written as

$$\hat{\nabla}_{\alpha_1}^2 u + \frac{\rho \omega^2}{\mu \alpha_2^2} u = 0, \tag{5.66}$$

where the operator $\hat{\nabla}_{\alpha_1}^2$ is defined by (5.49). It is also noted that $\alpha_1 \to R_1$ as $a \to 0$.

The domain with a small hole of radius a has a singularly perturbed boundary in the limit when $a \to 0$. Simultaneously, this is also a regularisation of the push-out transformation used to design an invisibility cloak.

Thus, we have the following boundary value problem

$$\left(\hat{\nabla}^2_{\alpha_1} + \frac{k^2}{\alpha_2^2}\right) u_1(r,\theta) = 0, \quad R_1 < r < R_2, \tag{5.67}$$

$$\left(\nabla^2 + k^2\right) u_2(R,\Theta) = 0, \quad R > R_2, \tag{5.68}$$

where $k = \omega\sqrt{\rho/\mu}$. The boundary conditions considered on the interior contour are the Dirichlet boundary condition (clamped boundary)

$$u_1(R_1,\theta) = 0, \tag{5.69}$$

or the Neumann boundary condition (free boundary)

$$\frac{\partial u_1}{\partial r}(R_1,\theta) = 0, \tag{5.70}$$

together with transmission conditions at the interface $r = R_2$

$$u_1(R_2,\theta) = u_2(R_2,\Theta), \tag{5.71}$$

$$\alpha_2 \frac{\partial u_1(R_2,\theta)}{\partial r} = \frac{\partial u_1(R_2,\theta)}{\partial R} = \frac{\partial u_2(R_2,\theta)}{\partial R}, \tag{5.72}$$

and the radiation condition as $R \to \infty$.

The solution of the Dirichlet problem (5.66)–(5.69), (5.71), and (5.72) may be written as

$$u_1(r,\theta) = \sum_{n=-\infty}^{\infty} \left[i^n J_n\left(\frac{k}{\alpha_2}(r-\alpha_1)\right) \right.$$

$$\left. - i^n \frac{J_n(ka)}{H_n^{(1)}(ka)} H_n^{(1)}\left(\frac{k}{\alpha_2}(r-\alpha_1)\right) \right] e^{in\theta} \tag{5.73}$$

and

$$u_2(R,\Theta) = \sum_{n=-\infty}^{\infty} \left[i^n J_n(kR) - i^n \frac{J_n(ka)}{H_n^{(1)}(ka)} H_n^{(1)}(kR) \right] e^{in\Theta}. \tag{5.74}$$

It is important to note that the outer solution $u_2(r,\theta)$ is the field of particular interest when examining the quality of cloaking. It is *exactly the same* as the field produced by a circular scatterer of radius a with Dirichlet boundary condition on the scatterer (Equations (5.54) and (5.56)). Thus the degree of disturbance of the incident plane wave is totally determined by the effect of this single uncloaked scatterer. In turn, this is then dependent on both the size of the circular scatterer and the boundary conditions on it.

The formulae (5.73) and (5.74) give the closed form analytical solution

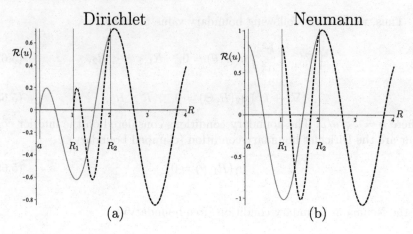

FIGURE 5.19: Membrane problem. Displacement $\mathcal{R}[u(R,0)]$ in the direction of the incident plane-wave propagation as a function of the distance R from the center of the scatterer or the cloaked region. Curves are given for the parameters $a = 0.1$ cm, $R_1 = 1$ cm, $R_2 = 2$ cm, corresponding to $\alpha_1 = 0.947$ and $\alpha_2 = 0.526$ cm, and $k = 3$ cm^{-1}. (a) Dirichlet problem: displacement in a homogeneous membrane with circular obstacle (grey continuous line, Equation (5.57)) and in the cloaking problem (black dashed line, Equations (5.73) and (5.74)). (b) Neumann problem: displacement in a homogeneous membrane with circular obstacle (grey continuous line, Equation (5.58)) and in the cloaking problem (black dashed line, Equations (5.75) and (5.76)). The sums in the multipole expansion representations of the displacements are truncated after the first 10 terms.

for a regularised invisibility cloak subject to the incident plane wave and the Dirichlet boundary condition on the interior contour of the cloaking layer.

In a similar way, when the boundary condition on the interior contour is the Neumann boundary condition (5.70), the formulae (5.73) and (5.74) are replaced by

$$u_1(r,\theta) = \sum_{n=-\infty}^{\infty} \left[i^n J_n\left(\frac{k}{\alpha_2}(r-\alpha_1)\right) \right.$$
$$\left. - i^n \frac{J_n'(ka)}{H_n^{(1)'}(ka)} H_n^{(1)}\left(\frac{k}{\alpha_2}(r-\alpha_1)\right) \right] e^{in\theta} \quad (5.75)$$

and

$$u_2(R,\Theta) = \sum_{n=-\infty}^{\infty} \left[i^n J_n(kR) - i^n \frac{J_n'(ka)}{H_n^{(1)'}(ka)} H_n^{(1)}(kR) \right] e^{in\Theta}. \quad (5.76)$$

The exterior field outside the cloaking layer is exactly the same as for the case of a small circular scatterer in Section 5.5.3.1 with either Dirichlet or

Neumann boundary conditions. The displacement fields for the homogeneous membrane with a circular scatterer and the cloaking problem are shown in Figure 5.19; they are given for the two types of boundary conditions. As $ka \to 0$ and according to (5.60) and (5.61) the quality of cloaking due to the cloaking layer with the Neumann boundary condition is higher than that due to the cloaking layer with the Dirichlet boundary condition. This is because the contributing coefficients in the scattered component of the solution are $O(|\log(ka)|^{-1})$ for the Dirichlet case and $O((ka)^2)$ for the Neumann case.

5.5.4 Singular perturbation and cloaking action for the biharmonic problem

Although it appears that both Dirichlet and Neumann boundary conditions for the membrane can be used for the design of the invisibility cloak, we have demonstrated that the logarithmic asymptotics in (5.60) and the power asymptotics of (5.61) and (5.62) suggest that the Neumann condition produces a more efficient cloak compared to the Dirichlet cloak as $a \to 0$.

This message concerning the boundary conditions is further reinforced by considering flexural waves. We will now consider the analytical solution to a further model problem when a plane flexural wave in a Kirchhoff–Love plate is incident on a small circular scatterer together with a corresponding cloaking problem in an analogous manner to that of Sections 5.5.3.1 and 5.5.3.2. Below, we will consider the biharmonic operator instead of the Helmholtz operator.

5.5.4.1 A model problem of scattering of a flexural wave by a circular scatterer

Consider an unbounded homogeneous isotropic Kirchhoff–Love plate with flexural rigidity D_0, thickness h and density ρ containing a clamped-edge hole of radius a. Further, incident time-harmonic elastic flexural plane waves of frequency ω result in a total scattered time-harmonic field of amplitude $u(\mathbf{X})$, at position $\mathbf{X}$. As in Section 5.5.3.1, cylindrical coordinates (R, Θ) will be used such that the direction of the incident plane waves is along the $\Theta = 0$ line. The field $u(R, \Theta)$ obeys the following problem outside the scatterer which is regarded as clamped,

$$(\Delta^2 - \beta^4)u(R, \Theta) = 0, \quad R > a; \tag{5.77}$$

$$u = 0, \quad \frac{\partial u}{\partial R} = 0, \quad R = a, \tag{5.78}$$

where $\beta^4 = \rho h \omega^2 / D_0$.

The problem is solved by factorising the operator into Helmholtz and modified Helmholtz operators in the usual way. Applying the radiation condition at infinity leads to the total field being composed of three contributions from the original plane wave, the Helmholtz and the modified Helmholtz fields. The general solution is

$$u(R,\Theta) = \sum_{n=-\infty}^{\infty} \left[i^n J_n(\beta R) + p_n H_n^{(1)}(\beta R) + q_n K_n(\beta R)\right] e^{in\Theta}, \quad (5.79)$$

where K_n is the modified Bessel function. Use has also been made of the plane wave expansion in cylindrical coordinates given in Equation (5.55). Application of the clamped boundary conditions in Equation (5.78) leads to the following expressions for the coefficients p_n and q_n

$$\begin{pmatrix} p_n \\ q_n \end{pmatrix} = \frac{1}{\mathcal{W}[H_n^{(1)}(\beta a), K_n(\beta a)]} \begin{pmatrix} K_n'(\beta a) & -K_n(\beta a) \\ -H_n^{(1)'}(\beta a) & H_n^{(1)}(\beta a) \end{pmatrix} \begin{pmatrix} -i^n J_n(\beta a) \\ -i^n J_n'(\beta a) \end{pmatrix}, \quad (5.80)$$

where

$$\mathcal{W}[H_n^{(1)}(\beta a), K_n(\beta a)] = H_n^{(1)}(\beta a) K_n'(\beta a) - H_n^{(1)'}(\beta a) K_n(\beta a).$$

For the case of free-edge boundary, the normal component of the moment and the transverse force are equal to zero on the boundary of the small circular scatterer, which results in the following conditions to be satisfied

$$\left[\frac{\partial^2 u}{\partial R^2} + \nu \left(\frac{1}{R}\frac{\partial u}{\partial R} + \frac{1}{R^2}\frac{\partial^2 u}{\partial \Theta^2}\right)\right] = 0, \quad \text{at} \quad R = a, \quad (5.81)$$

and

$$T_R(u) = 0, \quad \text{at} \quad R = a, \quad (5.82)$$

where

$$T_R(u) = -\left[\frac{\partial}{\partial R}\left(\frac{\partial^2 u}{\partial R^2} + \frac{1}{R}\frac{\partial u}{\partial R} + \frac{1}{R^2}\frac{\partial^2 u}{\partial \Theta^2}\right)\right] - \frac{1}{R}\frac{\partial M_{RT}(u)}{\partial \Theta}, \quad (5.83)$$

with

$$M_{RT}(u) = (1-\nu)\left[\frac{1}{R}\frac{\partial^2 u}{\partial R \partial \Theta} - \frac{1}{R^2}\frac{\partial u}{\partial \Theta}\right]. \quad (5.84)$$

Application of the free boundary conditions in Equation (5.78) instead of the clamped boundary conditions leads to the following expressions for the coefficients p_n and q_n

$$\begin{pmatrix} p_n \\ q_n \end{pmatrix} = \frac{1}{\mathscr{D}} \begin{pmatrix} V^-(K_n(\beta a)) & -X^+(K_n(\beta a)) \\ -V^+(H_n^{(1)}(\beta a)) & X^-(H_n^{(1)}(\beta a)) \end{pmatrix} \begin{pmatrix} -i^n X^-(J_n(\beta a)) \\ -i^n V^+(J_n(\beta a)) \end{pmatrix}, \quad (5.85)$$

where

$$X^{\pm}(\mathscr{C}_n(\beta a)) = -\beta a(1-\nu)\mathscr{C}_n'(\beta a) + [n^2(1-\nu) \pm (\beta a)^2]\mathscr{C}_n(\beta a), \quad (5.86)$$

$$V^{\pm}(\mathscr{C}_n(\beta a)) = [\pm(\beta a)^3 + (1-\nu)n^2 \beta a]\mathscr{C}_n'(\beta a) - (1-\nu)n^2 \mathscr{C}_n(\beta a), \quad (5.87)$$

with $\mathscr{C}_n(\beta a)$ denoting $H_n^{(1)}(\beta a)$ or $K_n(\beta a)$, and

$$\mathscr{D} = X^-(H_n^{(1)}(\beta a))V^-(K_n(\beta a)) - V^+(H_n^{(1)}(\beta a))X^+(K_n(\beta a)). \tag{5.88}$$

Compared to the membrane waves, additional action is produced by evanescent waves; in particular, Green's function for flexural waves remains bounded whereas for the membrane problem it is singular. We refer to the paper [89] that has addressed the classical solution of the scattering of a flexural plane wave. The first several multipole coefficients p_0 in the expansion (5.79) of the flexural displacement have the following asymptotics representations as $\beta a \ll 1$

$$p_0 = -1 + O((\beta a)^2), \quad p_1 = -\frac{\pi}{4\ln(\beta a)} + O((\beta a)^2), \tag{5.89}$$

$$p_2 = \frac{i\pi}{8}(\beta a)^2 + O((\beta a)^4), \quad p_3 = -\frac{\pi}{64}(\beta a)^4 + O((\beta a)^6),$$

for the case of the clamped boundary (Dirichlet problem), and

$$p_0 = \frac{i\pi\nu}{4(1-\nu)}(\beta a)^2 + \frac{i\pi(1+\nu)}{8(1-\nu)}\ln(\beta a)(\beta a)^4 + O((\beta a)^6), \tag{5.90}$$

$$p_1 = -\frac{\pi(1+\nu)}{32}(\beta a)^4 + O((\beta a)^6),$$

$$p_2 = -\frac{i\pi(1-\nu)}{8(3+\nu)}(\beta a)^2 + \frac{i\pi(6+2\nu-i\pi(1-\nu))}{64(3+\nu)^2}(\beta a)^4 + O((\beta a)^6),$$

$$p_3 = \frac{\pi(1-\nu)}{64(3+\nu)}(\beta a)^4 - \frac{\pi(1-\nu)}{384(3+\nu)}(\beta a)^6 + O((\beta a)^8),$$

for the case of the free-edge boundary (Neumann problem). We remark that the coefficients $p_n, |n| \geq 1$, vanish as $\beta a \to 0$, for both types of boundary conditions considered above. We also note that the coefficients q_n near K_n terms are not significant, as these terms are exponentially small away from the scatterer, as $\beta R \gg 1$.

For the Dirichlet problem, when the contour of the circular inclusion is clamped as in (5.78), the scattered field u_s is

$$u_s \sim -\sqrt{\frac{2}{\pi\beta R}} \exp\left(i\left(\beta R - \frac{\pi}{4}\right)\right) \quad \text{as } \beta R \to \infty. \tag{5.91}$$

For the case of a free-edge boundary condition, where both the transverse force and the moment vanish at $R = a$, the scattered field u_s becomes $O((\beta a)^2 \sqrt{\frac{2}{\pi\beta R}})$ as $\beta R \to \infty$. Equation (5.91) clearly illustrates that, even in the limit of $\beta a \to 0$, the scattered field produced by the "rigid pin" corresponds to a finite force. In other words, for the Dirichlet boundary condition, the singular perturbation of the boundary produces a finite scattered field and this remains finite in the neighbourhood of the scatterer even when the diameter of this scatterer tends to zero. On the contrary, for the case of the free-edge boundary, the contributing coefficients are $O((\beta a)^2)$, and the field vanishes as $\beta a \to 0$.

Comparison with (5.60) and (5.61) shows that, while for the Helmholtz operator, the monopole term in the scattered field always vanishes in the limit of $\beta a \to 0$ (both for Dirichlet and Neumann boundary conditions); in the biharmonic case, which corresponds to a Kirchhoff–Love plate, the scatterer with the clamped boundary (Dirichlet boundary conditions) delivers a finite non-zero point force in the limit of $\beta a \to 0$. On the other hand, the case of a free-edge boundary (Neumann boundary conditions) gives no scattering in the limit of $\beta a \to 0$.

5.5.4.2 Boundary conditions and the cloaking problem in a Kirchhoff–Love plate

The problem to be solved in this section is analogous to that in Section 5.5.3.2 except that we will consider cloaking of flexural rather than out-of-plane shear waves. A "near-cloak" is again defined through the invertible geometrical transformation of a plate with a small hole of radius a, into a plate containing a ring, with unperturbed exterior radius R_2 and an expanded interior radius R_1 as in Equations (5.63), (5.64) and (5.65).

Again, incident time-harmonic flexural waves of frequency ω result in a total scattered time-harmonic fields of amplitude $u_1(r, \theta)$ and $u_2(R, \Theta)$ inside and outside the cloak respectively. The material outside the ring is isotropic and homogeneous and the governing equation is as in Equation (5.77). Inside the cloak, the governing equation is given by the transformed biharmonic equation and the classical Laplace's operator is modified accordingly as in Equation (5.49). The problem to be solved is defined as

$$\left(\hat{\nabla}^4_{\alpha_1} - \frac{\beta^4}{\alpha_2^4}\right) u_1(r, \theta) = 0, \quad R_1 \leq r \leq R_2 \tag{5.92}$$

$$\left(\nabla^4 - \beta^4\right) u_2(R, \Theta) = 0, \quad R > R_2 \tag{5.93}$$

For the case of the clamped interior boundary, the boundary conditions are

$$u_1(R_1, \theta) = 0, \tag{5.94}$$

$$\frac{\partial u_1(R_1, \theta)}{\partial r} = 0. \tag{5.95}$$

For the case of free-edge boundary, zero transverse force and normal component of the moment on the interior contour of the cloaking region correspond to transformed boundary operators, obtained from the variational formulation for the shifted biharmonic operator in (5.92) used within the cloak. They result in the following conditions to be satisfied at $r = R_1$

$$\mathfrak{T}_r = -\alpha_2^3 \left[\frac{\partial}{\partial r}\left(\frac{\partial^2 u_1}{\partial r^2} + \frac{1}{(r-\alpha_1)}\frac{\partial u_1}{\partial r} + \frac{1}{(r-\alpha_1)^2}\frac{\partial^2 u_1}{\partial \theta^2}\right)\right]$$

$$-\frac{\alpha_2}{(r-\alpha_1)}\frac{\partial \mathfrak{M}_{rt}}{\partial \theta} = 0, \qquad (5.96)$$

with

$$\mathfrak{M}_{rt} = \alpha_2^2(1-\nu)\left[\frac{1}{(r-\alpha_1)}\frac{\partial^2 u_1}{\partial r \partial \theta} - \frac{1}{(r-\alpha_1)^2}\frac{\partial u_1}{\partial \theta}\right], \qquad (5.97)$$

and

$$\alpha_2^2\left[\frac{\partial^2 u_1}{\partial r^2} + \nu\left(\frac{1}{(r-\alpha_1)}\frac{\partial u_1}{\partial r} + \frac{1}{(r-\alpha_1)^2}\frac{\partial^2 u_1}{\partial \theta^2}\right)\right] = 0. \qquad (5.98)$$

In addition, on the interface between the cloak and the ambient material we have

$$u_1(R_2, \theta) = u_2(R_2, \Theta), \qquad (5.99)$$

$$\alpha_2 \frac{\partial u_1(R_2,\theta)}{\partial r} = \frac{\partial u_1(R_2,\theta)}{\partial R} = \frac{\partial u_2(R_2,\Theta)}{\partial R}, \qquad (5.100)$$

$$\left[\frac{\partial^2 u_2}{\partial R^2} + \nu\left(\frac{1}{R}\frac{\partial u_2}{\partial R} + \frac{1}{R^2}\frac{\partial^2 u_2}{\partial \Theta^2}\right)\right] = \alpha_2^2\left[\frac{\partial^2 u_1}{\partial r^2}\right.$$
$$\left. + \nu\left(\frac{1}{(r-\alpha_1)}\frac{\partial u_1}{\partial r} + \frac{1}{(r-\alpha_1)^2}\frac{\partial^2 u_1}{\partial \theta^2}\right)\right], \qquad (5.101)$$

for the normalised normal component of the moment on the outer boundary of the cloak, and

$$T_R(u_2) = \mathfrak{T}_r(u_1), \qquad (5.102)$$

for the normalised transverse force on the outer boundary of the cloak. The transverse force $T_R(u)$ is defined in (5.83) and (5.84).

We also supply the radiation condition as $R \to \infty$, i.e. the scattered field is represented as an outgoing wave.

For the homogeneous material, when $\alpha_2 = 1$ and $\alpha_1 = 0$, the above conditions imply the continuity of the flexural displacement and rotation, in addition to the normal moments and transverse forces.

The series representation of the flexural displacement field inside the cloak and outside the cloaking region has the form

$$u_1(r,\theta) = \sum_{n=-\infty}^{\infty}\left[a_n J_n\left(\frac{\beta}{\alpha_2}(r-\alpha_1)\right) + b_n H_n^{(1)}\left(\frac{\beta}{\alpha_2}(r-\alpha_1)\right)\right.$$
$$\left. + c_n I_n\left(\frac{\beta}{\alpha_2}(r-\alpha_1)\right) + d_n K_n\left(\frac{\beta}{\alpha_2}(r-\alpha_1)\right)\right]e^{in\theta}, \quad R_1 < r < R_2, \quad (5.103)$$

and

$$u_2(R,\Theta) = \sum_{n=-\infty}^{\infty}\left[i^n J_n(\beta R) + p_n H_n^{(1)}(\beta R) + q_n K_n(\beta R)\right]e^{in\Theta}, \quad R > R_2, \qquad (5.104)$$

respectively.

For each order n, there are six unknown coefficients. For the case of the clamped interior boundary $r = R_1$, a solution satisfying the six boundary and interface conditions (5.94)–(5.102) is given by $a_n = i^n$, $b_n = p_n$, $c_n = 0$, $d_n = q_n$ together with p_n and q_n satisfying

$$\begin{pmatrix} p_n \\ q_n \end{pmatrix} = \frac{1}{\mathcal{W}[H_n^{(1)}(\beta a), K_n(\beta a)]} \begin{pmatrix} K_n'(\beta a) & -K_n(\beta a) \\ -H_n^{(1)'}(\beta a) & H_n^{(1)}(\beta a) \end{pmatrix} \begin{pmatrix} -i^n J_n(\beta a) \\ -i^n J_n'(\beta a) \end{pmatrix},$$
(5.105)

where

$$\mathcal{W}[H_n^{(1)}(\beta a), K_n(\beta a)] = H_n^{(1)}(\beta a) K_n'(\beta a) - H_n^{(1)'}(\beta a) K_n(\beta a).$$

Note that these equations for p_n and q_n are the same as in (5.80) for the case of the clamped boundary $R = a$, and the corresponding coefficients for the case of the free-edge condition are the same as p_n and q_n in Equation (5.85), respectively.

This implies that the outer field $u_2(R, \Theta)$ for the cloaking problem is *exactly the same* as the field produced by a circular scatterer subjected to the same boundary conditions (either clamped or free-edge boundary) as those on the interior boundary of the cloak in the cloaking problem. The required solutions are given by Equations (5.79)–(5.88). The displacement fields are shown in Figure 5.20.

The asymptotic representation of the displacement field u_2, for $R \gg 1$, outside the regularised cloak, considered here, is also the same as the one for the small inclusion with either Dirichlet or Neumann boundary conditions, as in Section 5.5.4.1. In particular, according to the formulae (5.91), the cloaking action is non-existent if the interior contour of the cloaking layer in the Kirchhoff–Love plate is clamped (Dirichlet boundary condition set on $r = R_1$). On the other hand, the cloaking appears to be efficient for the free-edge boundary, with the leading term of the scattered field to be of order $(\beta a)^2$ as $\beta a \to 0$.

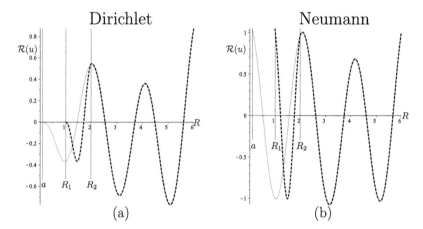

FIGURE 5.20: Plate problem. Displacement $\mathcal{R}[u(R,0)]$ in the direction of the incident plane-wave propagation as a function of the distance R from the center of the scatterer or the cloaked region. Curves are given for the parameters $a = 0.1$ cm, $R_1 = 1$ cm, $R_2 = 2$ cm, corresponding to $\alpha_1 = 0.947$ and $\alpha_2 = 0.526$ cm, $\beta = 3$ cm^{-1} and $\nu = 0.3$. (a) Dirichlet problem: displacement in a homogeneous plate with circular obstacle (grey continuous line, Equations (5.79) and (5.80)) and in the cloaking problem (black dashed line, Equations (5.103)–(5.105)). (b) Neumann problem: displacement in a homogeneous plate with circular obstacle (grey continuous line, Equations (5.79) and (5.85)) and in the cloaking problem (black dashed line). The sums in the multipole expansion representations of the displacements are truncated after the first 10 terms.

Chapter 6

Structured interfaces and chiral systems in dynamics of elastic solids

As we have already seen, two geometrically identical lattices with the same elastic stiffness but different inertia may provide different dynamic responses. If two such lattices were merged, in statics one would have a uniform homogeneous lattice, but its dynamic properties may show blockage of waves as well as localisation, filtering and wave polarisation.

The term "structured interface" is used for an interface region (continuum or discrete), which possesses a micro-structure and separates two parts of a solid or a lattice. Structured interfaces may possess inertia, and tractions may become discontinuous across such interfaces in dynamic regimes. The derivation and analysis of dynamic transmission conditions for structured interfaces are important for the evaluation of transmission and reflection properties of heterogeneous composite media. In addition, transmission resonances for elastic waves, as well as negative refraction and focusing are of great interest.

This chapter is based on the papers [30, 31, 49], and it addresses shear polarisation of waves, modelling of active chiral media and dynamic response of discrete structured interfaces, which may exhibit shielding, negative refraction and focussing of elastic waves.

6.1 Structured interface as a polarising filter

Here we consider a shear-type structured interface excited by a plane pressure wave at an oblique incidence. A stratified elastic structure is embedded into a horizontal interface, and it is assumed to have a low effective shear stiffness. The vertical (transverse) displacements within the interface are assumed to be negligibly small.

The model discussed in this section is based on the results of the paper [31]; the physical configuration of the system under consideration is also illustrated in Figure 6.1. In this case, the interface is composed of horizontally aligned elastic layers, connected by transverse elastic bars whose flexural rigidity is small compared to their longitudinal stiffness.

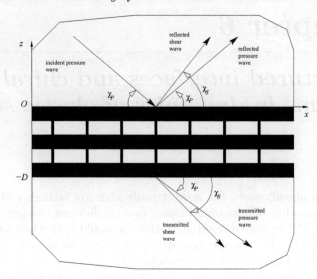

FIGURE 6.1: Two elastic half-planes are joined by the structured interface. The structured interface consists of horizontal elastic bars, which can move in the x-direction, while the vertical displacements are negligible.

6.1.1 Stratified domain

Consider an elastic region incorporating a structured interface of finite width D. The interface region is a horizontal strip in the (x, z)-plane:

$$\Pi_D = \{(x, z) : x \in \mathbb{R}, -D \leq z \leq 0\}. \tag{6.1}$$

The structure within the interface incorporates N elastic longitudinal bars l_j, $j = 1, 2, ..., N$, connected to each other via transverse elastic massless links. Ideal elastic contact, implying continuity of tractions and displacements, is imposed at the upper boundary of the interface as well as at its lower boundary. The notations Ω_+ and Ω_- are used for the half-planes above and below the structured interface, respectively:

$$\Omega_+ = \{(x, z) : x \in \mathbb{R}, z > 0\}, \quad \Omega_- = \{(x, z) : x \in \mathbb{R}, z < -D\}.$$

We describe the structure of the interface within Π_D in the next section.

6.1.2 Lower-dimensional approximations within the interface

Since the interface consists of thin structured elements, the actual elastic material within the interface is assumed to be sufficiently stiff, as described in the text below. Each elastic bar is treated as a one-dimensional elastic element, and we assume that the leading approximation of the elastic displacement field within the bar l_j has the form

$$\mathbf{u}^{(j)} \sim u_j(x, t)\mathbf{e}^{(1)}, \tag{6.2}$$

where $\mathbf{e}^{(1)}$ is the basis Cartesian vector in the direction of the x-axis. The approximation (6.2) implies that each bar moves only in the horizontal direction (along the x-axis). Such a state can be achieved by assuming that the upper horizontal elastic bar is constrained against any motion in the vertical direction, along the z-axis, and that the horizontal elastic bars have a very high bending stiffness.

The equations of motion can be written for the elastic bars inside the interface ($j = 2, 3, \ldots, N-1$) as well as for those on the boundary of the interface adjacent to Ω_+ and Ω_- ($j = 1$ and $j = N$). Let l_j be an interior elastic bar, $1 < j < N$, and let E denote the longitudinal stiffness, ρ the mass density, and γ the shear stiffness of the array of vertical elastic links connecting l_j with l_{j-1} and l_{j+1}. The equation of motion for the interior bar l_j takes the form:

$$E(u_j)_{xx} - \rho(u_j)_{tt} - 2\gamma u_j + \gamma(u_{j+1} - u_{j-1}) = 0, \quad j = 2, 3, \ldots, N-1, \quad (6.3)$$

where $(u_j)_{xx}$ and $(u_j)_{tt}$ stand for the second-order partial derivatives with respect to x and t, respectively.

For the bars l_1 and l_N on the boundary of the structured interface, the equations of motion are

$$E(u_1)_{xx} - \rho(u_1)_{tt} + \gamma(u_2 - u_1) + \tau_+ = 0, \quad (6.4)$$

$$E(u_N)_{xx} - \rho(u_N)_{tt} + \gamma(u_{N-1} - u_N) - \tau_- = 0, \quad (6.5)$$

where τ_+ and τ_- are the shear stresses in the ambient elastic media above and below the interface.

6.1.3 Incident, reflected and transmitted waves

We assume that a plane pressure wave in Ω_+ is incident, at a certain angle, on the structured interface. Both reflected pressure and shear plane waves are generated in Ω_+, while the transmitted pressure and shear waves occur in Ω_-. Conventionally, we represent the displacement field in the elastic continuum above and below the structured interface

$$\mathbf{u} = u(x, z, t)\mathbf{e}^{(1)} + w(x, z, t)\mathbf{e}^{(3)}$$

via the pressure and shear wave potentials $\varphi(x, z, t)$ and $\psi(x, z, t)\,\mathbf{e}^{(2)}$, so that the elastic displacement has the form $\nabla\varphi + \nabla \times (\psi\,\mathbf{e}^{(2)})$, and

$$u = \frac{\partial \varphi}{\partial x} - \frac{\partial \psi}{\partial z}, \quad w = \frac{\partial \varphi}{\partial z} + \frac{\partial \psi}{\partial x}. \quad (6.6)$$

We shall use the notations $\{\varphi^{(I)}, \psi^{(I)}\}$, $\{\varphi^{(R)}, \psi^{(R)}\}$, $\{\varphi^{(T)}, \psi^{(T)}\}$ for the potentials corresponding to the incident, reflected and transmitted waves, respectively. Note that, since the incident wave is of pressure type, $\psi^{(I)} \equiv 0$.

Let c be an apparent velocity of the incident wave along the horizontal interface, and let α, β be the wave speeds of the pressure and shear waves in $\Omega_\pm$, so that

$$\Delta\varphi - \frac{1}{\alpha^2}\frac{\partial^2}{\partial t^2}\varphi = 0, \qquad \Delta\psi - \frac{1}{\beta^2}\frac{\partial^2}{\partial t^2}\psi = 0, \qquad (6.7)$$

where

$$\alpha = \sqrt{\frac{\lambda + 2\mu}{\rho}}, \qquad \beta = \sqrt{\frac{\mu}{\rho}}. \qquad (6.8)$$

If $\chi_P \in [0, \pi/2]$ denotes the angle between the direction of the incident wave and the horizontal interface, then

$$a := \tan\chi_P = \sqrt{\frac{c^2}{\alpha^2} - 1}. \qquad (6.9)$$

For the pressure wave, the angle of reflection is equal to the angle of incidence, while for the reflected shear wave, the angle $\chi_S \in [0, \pi/2]$ between the direction of propagation and the horizontal interface is defined by

$$b := \tan\chi_S = \sqrt{\frac{c^2}{\beta^2} - 1}. \qquad (6.10)$$

Note that $\chi_S > \chi_P$; the two angles are shown in Figure 6.1.

The representations for the potentials in Ω_+ are given by

$$\begin{aligned} \varphi_+ &= \varphi^{(I)} + \varphi^{(R)} = A_I \exp[ik(ct - x + az)] + A_R \exp[ik(ct - x - az)], \\ \psi_+ &= \psi^{(R)} = B_R \exp[ik(ct - x - bz)], \end{aligned} \qquad (6.11)$$

while in Ω_-

$$\begin{aligned} \varphi_- &= \varphi^{(T)} = A_T \exp[ik(ct - x + az)], \\ \psi_- &= \psi^{(T)} = B_T \exp[ik(ct - x + bz)]. \end{aligned} \qquad (6.12)$$

In (6.11) and (6.12)

$$k = \frac{2\pi}{\Lambda_P}\cos\chi_P = \frac{2\pi}{\Lambda_S}\cos\chi_S, \qquad (6.13)$$

where Λ_P and Λ_S are the wavelengths of the pressure and shear waves in the ambient elastic medium $\Omega_+ \cup \Omega_-$. We shall also use the notation $\omega = kc$ for the radian frequency.

Assuming that the amplitude A_I is given, we will evaluate A_R, B_R, A_T, B_T, and show that the shear wave may become dominant in the region of transmission Ω_-, so that the structured interface will act as a shear polariser.

6.1.4 The energy of transmitted and reflected waves

Assuming that the motion is time-harmonic with the radian frequency ω, we use the notations U_j and $T_\pm$ for the amplitudes of the displacements of the bars $l_j, j = 1, 2, 3$ and for the shear stresses above and below the interface.

Then the equations of motion of the elastic bars within the shear-type interface lead to the algebraic system

$$\mathcal{R}\mathbf{U} = \mathbf{T}, \tag{6.14}$$

where

$$\mathbf{U} = (U_1, U_2, U_3)^T, \ \mathbf{T} = (-T_+, 0, T_-)^T, \tag{6.15}$$

and

$$\mathcal{R} = \begin{pmatrix} -\Gamma & \gamma & 0 \\ \gamma & -\Gamma - \gamma & \gamma \\ 0 & \gamma & -\Gamma \end{pmatrix}, \tag{6.16}$$

with $\Gamma = k^2 E + \gamma - \rho\omega^2$.

As described in [31] the procedure for deriving this linear system involves the Betti formula applied to the displacement and its complex conjugate above and below the interface. Furthermore, using the plane wave representations (6.11) and (6.12) that include the pressure and shear wave potentials, we deduce the following energy balance relation

$$E_I = E_R + E_T, \tag{6.17}$$

where

$$E_I = a|A_I|^2, E_R = a|A_R|^2 + b|B_R|^2, E_T = a|A_T|^2 + b|B_T|^2, \tag{6.18}$$

E_I, E_R and E_T represent the vertical energy fluxes of the incident, reflected and transmitted fields, respectively.

The effect of the interface on the distribution of energy can be seen by the evaluation of the coefficients A_R, B_R and A_T, B_T characterising the reflected and transmitted fields.

6.1.5 Trapped waveforms

Trapped vibrations within the structured interface can significantly alter its transmission properties. In the particular case involving the three-bar interface we discuss several examples in this section.

For a non-resonant regime, when $\det \mathcal{R} \neq 0$, the system (6.14) has a solution

$$\begin{pmatrix} U_1 \\ U_2 \\ U_3 \end{pmatrix} = \mathcal{R}^{-1} \begin{pmatrix} -T_+ \\ 0 \\ T_- \end{pmatrix}, \tag{6.19}$$

where

$$\mathcal{R}^{-1} = \frac{1}{(\gamma-\Gamma)(2\gamma+\Gamma)} \begin{pmatrix} \gamma - \frac{\gamma^2}{\Gamma} + \Gamma & \gamma & \frac{\gamma^2}{\Gamma} \\ \gamma & \Gamma & \gamma \\ \frac{\gamma^2}{\Gamma} & \gamma & \gamma - \frac{\gamma^2}{\Gamma} + \Gamma \end{pmatrix}. \quad (6.20)$$

In particular, the effective transmission relations for the shear displacements and shear stresses across the structured interface have the form:

$$U_1 = \frac{\frac{\gamma^2}{\Gamma}(T_+ + T_-) - T_+(\gamma + \Gamma)}{(\gamma - \Gamma)(2\gamma + \Gamma)},$$

$$U_3 = \frac{T_-(\gamma + \Gamma) - \frac{\gamma^2}{\Gamma}(T_+ + T_-)}{(\gamma - \Gamma)(2\gamma + \Gamma)}. \quad (6.21)$$

Solving the above system of algebraic equations with respect to B_R, B_T we deduce that

$$B_R = -\frac{2a}{\Psi}\left[\Phi(\Gamma^2 + \gamma\Gamma - \gamma^2) - \gamma^2\Gamma(1+ab)\right]A_I,$$

$$B_T = \frac{2ia^2(1+b^2)}{\Psi}k\gamma^2\mu e^{ibDk}A_I, \quad (6.22)$$

where $\Phi = \Gamma(1+ab) + ik\mu a(1+b^2)$ and $\Psi = \Phi\left[(\gamma+\Gamma)\Phi - 2\gamma^2(1+ab)\right]$. The conditions of zero transverse displacements at the interface lead to the expressions for the coefficients A_R, A_T as follows

$$A_R = A_I - a^{-1}B_R, \qquad A_T = a^{-1}B_T e^{-ik(b-a)D}, \quad (6.23)$$

The full set $\{A_R, B_R, A_T, B_T\}$ determines the reflected and transmitted waves at the shear-type interface.

Furthermore, the system (6.14) can be written as follows

$$\mathbf{B}\begin{pmatrix} [\![.]\!]U \\ [\![.]\!]V \\ \langle U \rangle \end{pmatrix} = \begin{pmatrix} T_+ + T_- \\ \frac{1}{2}(T_- - T_+) \\ \frac{1}{3}(T_- - T_+) \end{pmatrix}, \quad (6.24)$$

where $[\![.]\!]U = U_3 - U_1$ represents the tangential displacement jump across the interface, $[\![.]\!]V = \frac{1}{2}(U_3 + U_1) - U_2$ is the average jump representing the difference between the average tangential displacement on the boundary and the tangential displacement of the interior bar, $\langle U \rangle = \frac{1}{3}(U_1 + U_2 + U_3)$ is the average tangential displacement over the whole structured interface, and

$$\mathbf{B} = \text{diag}\left\{-Ek^2 + \rho\omega^2 - \gamma, -Ek^2 + \rho\omega^2 - 3\gamma, -Ek^2 + \rho\omega^2\right\}. \quad (6.25)$$

The relation between $([\![.]\!]U, [\![.]\!]V, \langle U \rangle)^T$ and $(U_1, U_2, U_3)^T$ has the form

$$\begin{pmatrix} [\![.]\!]U \\ [\![.]\!]V \\ \langle U \rangle \end{pmatrix} = \mathcal{Q}\begin{pmatrix} U_1 \\ U_2 \\ U_3 \end{pmatrix}, \quad (6.26)$$

where
$$\mathcal{Q} = \begin{pmatrix} -1 & 0 & 1 \\ \frac{1}{2} & -1 & \frac{1}{2} \\ \frac{1}{3} & \frac{1}{3} & \frac{1}{3} \end{pmatrix}, \qquad (6.27)$$

and hence
$$\mathbf{B} = \mathcal{Q}\mathcal{R}\mathcal{Q}^{-1}. \qquad (6.28)$$

The columns of the matrix $\mathcal{Q}$ are the eigenvectors of $\mathcal{R}$ corresponding to the eigenvalues, which coincide with the diagonal entries of the matrix $\mathbf{B}$, and
$$\det \mathbf{B} = \det \mathcal{R}. \qquad (6.29)$$

The relation
$$\det \mathcal{R}(\omega, k) = 0 \qquad (6.30)$$
is the dispersion equation for the elastic waves propagating horizontally along the structured interface, which is equivalent to
$$(-Ek^2 + \rho\omega^2 - \gamma)(-Ek^2 + \rho\omega^2 - 3\gamma)(-Ek^2 + \rho\omega^2) = 0. \qquad (6.31)$$

The corresponding dispersion branches are defined by the equations:
$$\begin{aligned}
(1) &\quad Ek^2 + 3\gamma - \rho\omega^2 = 0, \\
(2) &\quad Ek^2 + \gamma - \rho\omega^2 = 0, \\
(3) &\quad Ek^2 - \rho\omega^2 = 0.
\end{aligned} \qquad (6.32)$$

The lowest branch (3) gives a linear relation between k and ω, which corresponds to a non-dispersive wave propagating along an elastic bar. For the dispersive waves, governed by branches (1) and (2), there is a cut-off frequency
$$\omega^* = \sqrt{\frac{\gamma}{\rho}}, \qquad (6.33)$$
and for $\omega < \omega^*$ no dispersive wave can propagate along the interface.

The resonance states correspond to the cases where the frequency ω of the incident wave coincides with one of the solutions of the dispersion equation (6.31). For the three-bar interface, such states can be classified as follows:

(a) When $Ek^2 - \rho\omega^2 = 0$, the compatibility condition for the system (6.24) is $T_- = T_+ = T$, which yields
$$[.]V = 0, \qquad [.]U = -\frac{2T}{\gamma}. \qquad (6.34)$$

(b) If $Ek^2 + 3\gamma - \rho\omega^2 = 0$, Equations (6.24) have a solution when $T_- = T_+ = T$, and hence
$$\langle U \rangle = 0, \qquad [.]U = \frac{T}{\gamma}. \qquad (6.35)$$

(c) In turn, for the case when $Ek^2+\gamma-\rho\omega^2 = 0$, the compatibility constraint is $-T_- = T_+ = T$, which leads to

$$\langle U \rangle = -\frac{2T}{3\gamma}, \quad [.]V = \frac{T}{2\gamma} \quad \text{and} \quad U_2 = -\frac{T}{\gamma}. \tag{6.36}$$

The amplitudes of the reflected and transmitted shear waves B_R, B_T in the above cases (a), (b), (c) are

$$\text{(a)} \quad B_R = \frac{a\gamma}{\gamma(1+ab) + ia(1+b^2)k\mu} A_I, \quad B_T = B_R e^{ibDk},$$

$$\text{(b)} \quad B_R = \frac{2a\gamma}{2\gamma(1+ab) - ia(1+b^2)k\mu} A_I, \quad B_T = B_R e^{ibDk},$$

$$\text{(c)} \quad B_R = \frac{2a\gamma}{2\gamma(1+ab) - ia(1+b^2)k\mu} A_I, \quad B_T = -B_R e^{ibDk}. \tag{6.37}$$

We note that in all above cases $|B_R| = |B_T|$. Hence we deduce that for the special resonance modes described above, the energy of the transmitted shear wave is equal to the energy of the reflected shear wave.

6.1.6 Enhanced transmission

The Dispersion of waves in media with boundaries or built-in microstructure is a fundamental phenomenon, which occurs in problems of models of water waves, simple systems involving one-dimensional harmonic oscillators, as well as complex elastic systems leading to analysis of full vector problems of elasticity both for discrete and continuous systems. If the system is infinite and periodic, the analysis is reduced to an elementary cell and dispersive properties of Bloch–Floquet waves can be represented via dispersion diagrams. In such problems, stop bands may exist and hence one can identify intervals of frequencies, within which the waves become evanescent and hence are confined exponentially. Here we draw a connection between the Bloch–Floquet waves and transmission across structured interfaces.

In Figure 6.2 we show the graphs of the reflected and transmitted energies as functions of c and k. In particular, the diagrams (a) and (c) show the surface plots of $E_R(k, c/\alpha)$ and $E_T(k, c/\alpha)$, whereas the parts (b) and (d) include the corresponding contour plots. We note that there is a rapid increase in the energy of transmission when k is close to the value $\sqrt{2\gamma/(\rho c^2 - E)}$ for $c > \sqrt{E/\rho}$. This special case corresponds to a *defect resonance mode*, for which the upper and lower bars within the interface do not move, whereas the middle bar vibrates while being connected to the upper and lower bars via elastic links of stiffness γ. The corresponding equation of motion for such a defect mode leads to

$$(Ek^2 + 2\gamma - \rho\omega^2)U_2 = 0, \tag{6.38}$$

which is fully consistent with the observed peak in transmission.

The computations in Figure 6.2 are based on Equations (6.23), (6.22), (6.37) and (6.18).

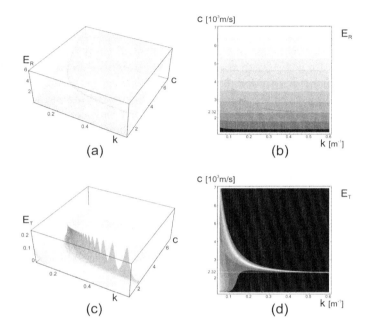

FIGURE 6.2: The reflected and transmitted energies E_R and E_T for a three-bar interface as functions of the wave number k and the apparent velocity c. Diagrams (a) and (c) show the surface plots of E_R and E_T. Diagrams (b) and (d) present the corresponding contour plots. The material parameters for the interface are chosen as follows. The bars have the linear mass density $\rho = 1000$ kg/m, and the longitudinal stiffness $E = 5 \times 10^6$ kN; the horizontal stiffness of the vertical links is $\gamma = 0.05$ GPa. The ambient medium is of the mass density $\rho_{\mathrm{amb}} = 1000$ kg$\cdot$m^{-3}, Poisson's ratio $\nu_{\mathrm{amb}} = 0.3$ and Young's modulus $E_{\mathrm{amb}} = 1$ GPa.

6.2 Vortex-type resonators and chiral polarisers of elastic waves

This section is based on the work [30], which introduced the notion of an active gyro-lattice containing appropriately tuned spinners. The precession of gyroscopes leads to formation of vortex-like waveforms, which impacts upon the polarisation of elastic waves in such a structured medium.

A general algorithm for analysing the spectral properties of a periodic multi-scale elastic lattice with resonators was established in [108], and lattice systems with built-in dynamic rotational interactions were analysed in [30]. One important feature of lattices with vortex-type resonators, periodically embedded into a lattice, is their influence on the dispersion of elastic waves and a special type of coupling between waves of different polarisations.

6.2.1 Governing equations

We consider a triangular periodic lattice with the gyroscopic resonators, as shown in Figure 6.3. The two-dimensional elastic lattice includes massless connecting ligaments, and point masses placed at the nodal points. The nodal points of the lattice are connected to the gyroscopic spinners, as illustrated in Figures 6.3a,c. The main triangular lattice is defined through the basis vectors:

$$\mathbf{t}^{(1)} = (2l, 0)^T \text{ and } \mathbf{t}^{(2)} = (l/2, l\sqrt{3}/2)^T. \tag{6.39}$$

In this case, we have two junctions per elementary cell of the periodic structure, and it is assumed that the point masses placed at these junctions are m_1 and m_2. The stiffnesses of elastic truss-like links, each of length l, are assumed to be equal to c, and the mass density of the elastic links is assumed to be negligibly small.

We also use the three unit vectors $\mathbf{a}^{(j)}, j = 1, 2, 3$, characterising the directions of trusses (see Figure 6.3b) within the elementary cell:

$$\mathbf{a}^{(1)} = (1,0)^T, \ \mathbf{a}^{(2)} = (-1/2, \sqrt{3}/2)^T, \ \mathbf{a}^{(3)} = (-1/2, -\sqrt{3}/2)^T.$$

The multi-index $\mathbf{n} = (n_1, n_2)$ with integer entries is used to characterise the position of the elementary cell within the periodic structure, so that the position vector of the mass m_j ($j = 1, 2$) is defined as

$$\mathbf{x}^{(\mathbf{n},j)} = \mathbf{x}^{(\mathbf{0},j)} + n_1 \mathbf{t}^{(1)} + n_2 \mathbf{t}^{(2)}, \tag{6.40}$$

where the vectors $\mathbf{t}^{(1)}$ and $\mathbf{t}^{(2)}$ are given in (6.39) and the displacement $\mathbf{u}^{(\mathbf{n},\kappa)} e^{i\omega t}$ is time-harmonic, of radian frequency ω. For convenience, we also use the multi-indices $\mathbf{e}_1 = (1, 0)$ and $\mathbf{e}_2 = (0, 1)$.

The design of the lattice system considered here includes gyroscopic spinners attached to every junction. The axis of each spinner is perpendicular to

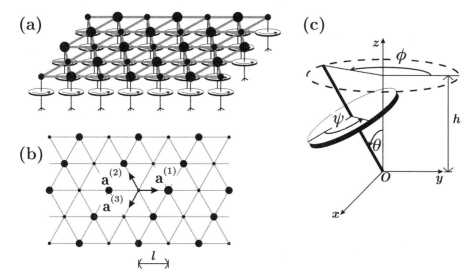

FIGURE 6.3: A triangular lattice linked to the system of spinners is shown in part (a). Part (b) shows the geometry of the triangular biatomic lattice. Different masses at junction points are shown as large and small discs. Part (c) shows a spinner and the notations for the coordinate systems; ψ, ϕ and θ are the angles of spin, precession and nutation, respectively.

the plane of the two-dimensional lattice, and its angular velocity is related to the radian frequency ω of the oscillation of the main lattice system. A small in-plane displacement of a junction leads to a change of orientation of the spinner axis and hence generates a moment creating a "vortex-type" effect.

The equations of motion for the nodal points equipped with masses m_1 and m_2 within the macro-cell containing resonators have the form derived in [108], but they also include gyroscopic terms where the gyroscopic force is orthogonal to the displacement, and it also features the phase shift, as discussed in [30]. Using the notations α_1, α_2 for the spinner constants, and $\mathbf{R}$ for the rotation matrix

$$\mathbf{R} = \begin{pmatrix} 0 & 1 \\ -1 & 0 \end{pmatrix},$$

we write the equations of motion in the form

$$-\frac{m_1\omega^2}{c}\mathbf{u}^{(n,1)} = \mathbf{a}^{(1)} \cdot (\mathbf{u}^{(n,2)} - \mathbf{u}^{(n,1)})\mathbf{a}^{(1)} + (-\mathbf{a}^{(1)}) \cdot (\mathbf{u}^{(n-e_1,2)} - \mathbf{u}^{(n,1)})(-\mathbf{a}^{(1)})$$

$$+ \mathbf{a}^{(2)} \cdot (\mathbf{u}^{(n-e_1+e_2,2)} - \mathbf{u}^{(n,1)})\mathbf{a}^{(2)} + (-\mathbf{a}^{(2)}) \cdot (\mathbf{u}^{(n-e_2,2)} - \mathbf{u}^{(n,1)})(-\mathbf{a}^{(2)})$$

$$+ \mathbf{a}^{(3)} \cdot (\mathbf{u}^{(n-e_2,1)} - \mathbf{u}^{(n,1)})\mathbf{a}^{(3)} + (-\mathbf{a}^{(3)}) \cdot (\mathbf{u}^{(n+e_2,1)} - \mathbf{u}^{(n,1)})(-\mathbf{a}^{(3)})$$

$$+ \frac{i\alpha_1\omega^2}{c}\mathbf{R}\mathbf{u}^{(n,1)}, \qquad (6.41)$$

and

$$-\frac{m_2\omega^2}{c}\mathbf{u}^{(n,2)} = \mathbf{a}^{(1)} \cdot (\mathbf{u}^{(n+e_1,1)} - \mathbf{u}^{(n,2)})\mathbf{a}^{(1)} + (-\mathbf{a}^{(1)}) \cdot (\mathbf{u}^{(n,1)} - \mathbf{u}^{(n,2)})(-\mathbf{a}^{(1)})$$
$$+\mathbf{a}^{(2)} \cdot (\mathbf{u}^{(n+e_2,1)} - \mathbf{u}^{(n,2)})\mathbf{a}^{(2)} + (-\mathbf{a}^{(2)}) \cdot (\mathbf{u}^{(n+e_1-e_2,1)} - \mathbf{u}^{(n,2)})(-\mathbf{a}^{(2)})$$
$$+\mathbf{a}^{(3)} \cdot (\mathbf{u}^{(n-e_2,2)} - \mathbf{u}^{(n,2)})\mathbf{a}^{(3)} + (-\mathbf{a}^{(3)}) \cdot (\mathbf{u}^{(n+e_2,2)} - \mathbf{u}^{(n,2)})(-\mathbf{a}^{(3)})$$
$$+\frac{i\alpha_2\omega^2}{c}\mathbf{R}\mathbf{u}^{(n,2)}. \tag{6.42}$$

6.2.2 Evaluation of the spinner constants

It is assumed that each junction within the lattice is connected to a spinner, whose axis is perpendicular to the plane of the lattice, as shown in Figure 6.3. By considering one of the junctions, together with the spinner attached, we evaluate the spinner constants $\alpha_j, j = 1, 2$, the real coefficients which have dimension of mass.

Let ψ, ϕ and θ be the angles of spin, precession and nutation of a spinner with respect to the vertical axis Oz, as shown in Figure 6.3c. Then, assuming that the gravity force is absent, and following [72, page 210], the equations of motion of the spinner, for the case of the constant spin rate ($\ddot{\psi} = 0$), can be written as follows

$$M_x = I_0(\ddot{\theta} - \dot{\phi}^2 \sin\theta \cos\theta) + I\dot{\phi}\sin\theta(\dot{\phi}\cos\theta + \dot{\psi}), \tag{6.43}$$

$$M_y = \sin\theta\Big(I_0(\ddot{\phi}\sin\theta + 2\dot{\phi}\dot{\theta}\cos\theta) - I\dot{\theta}(\dot{\phi}\cos\theta + \dot{\psi})\Big), \tag{6.44}$$

$$M_z = I(\ddot{\phi}\cos\theta - \dot{\phi}\dot{\theta}\sin\theta), \tag{6.45}$$

where M_x, M_y, M_z are the moments about the x, y and z axes, and $I_0 = I_{xx} = I_{yy}, I = I_{zz}$ are the moments of inertia.

We consider a small amplitude time harmonic motion (due to the motion of the lattice attached to the spinner), resulting in the nutation angle

$$\theta(t) = \Theta e^{i\omega t}, \quad |\Theta| \ll 1,$$

and constant spin rate $\dot{\psi} = \Omega$.

Assuming that there are no imposed moments about the x and y axes (i.e. $M_x = M_y = 0$), and the precession rate is constant ($\ddot{\phi} = 0$), we deduce the connection between the spin rate Ω and the radian fequency ω of the imposed nutation. Namely, the precession rate is given as

$$\dot{\phi} = \frac{I\Omega}{2I_0 - I}. \tag{6.46}$$

According to the equation of motion (6.43) we have

$$(I - I_0)\dot{\phi}^2 + I\Omega\dot{\phi} - I_0\omega^2 = 0. \tag{6.47}$$

The compatibility condition for Equations (6.46) and (6.47) has the form

$$\Omega = \pm\omega\frac{2I_0 - I}{I}. \tag{6.48}$$

Direct substitution into (6.45) gives the induced moment about the z axis

$$M_z = \mp I\omega\dot{\theta}\theta.$$

Let h be the characteristic length of the spinner, as shown in Figure 6.3c. Then the magnitude of the in-plane displacement in the lattice junction is

$$U = \theta h.$$

Taking into account that the moment about the z axis is $M_z = Fh\theta$, we deduce that the "rotational force" F is given by

$$F = \mp Ii\frac{\omega^2}{h^2}U,$$

which indicates the presence of a phase shift in the gyroscopic force compared to the time-harmonic displacement. Following the above derivation, the spinner constants α_1 and α_2 in Equations (6.41) and (6.42) are

$$\alpha_j = h^{-2}I_{zz}^{(j)}, \ j = 1, 2.$$

6.2.3 Elastic Bloch–Floquet waves in the active chiral lattice

Consider the case when $m_1 = m_2$ and $\alpha_1 = \alpha_2$, i.e. the monatomic lattice. Then, Equations (6.41) and (6.42) are equivalent, subject to a translation of an elementary cell of the periodic system, which includes only a single junction. This configuration is a good example for observing the influence of the vortex-type interaction within the lattice on the dispersive properties of the elastic waves. The amplitudes U_1, U_2 of the time-harmonic displacement, of radian frequency ω, within the periodic vortex-type lattice satisfy the system of equations (6.41) together with the Bloch–Floquet conditions set within the elementary cell of the periodic system, as follows:

$$\mathbf{u}(\mathbf{x} + n_1\mathbf{t}^{(1)}/2 + n_2\mathbf{t}^{(2)}) = \mathbf{u}(\mathbf{x})\exp(i\mathbf{k}\cdot\mathbf{Tn}), \tag{6.49}$$

where $\mathbf{k}$ is the Bloch vector $(k_1, k_2)^T$, and the matrix $\mathbf{T}$ has the form

$$\mathbf{T} = \begin{pmatrix} l & l/2 \\ 0 & l\sqrt{3}/2 \end{pmatrix}.$$

The formula (6.49) refers to the case of a monatomic lattice where the elementary cell is smaller compared to the case of a biatomic system (compare with (6.40)). These waves are dispersive, and the corresponding dispersion equation, as in [30], is

$$\det\left[\mathbf{C}(\mathbf{k}) - \omega^2(\mathbf{M} - \boldsymbol{\Sigma})\right] = 0, \tag{6.50}$$

where the mass matrix $\mathbf{M} = \mathrm{diag}[m, m]$ and $\boldsymbol{\Sigma}$ represents the chiral term associated with the presence of spinners attached to the junction points of the lattice system. Assuming that each spinner is characterised by the "spinner constant" α, we represent the matrix $\boldsymbol{\Sigma}$ as

$$\boldsymbol{\Sigma} = \begin{pmatrix} 0 & -i\alpha \\ i\alpha & 0 \end{pmatrix}. \tag{6.51}$$

Additionally, in Equation (6.50), $\mathbf{C}(\mathbf{k})$ is the stiffness matrix for the monatomic lattice of masses m, with spring connectors of stiffness c. Similar to [108], it is given by

$$\mathbf{C}(\mathbf{k}) = c \begin{pmatrix} 3 - 2\cos k_1 l - \dfrac{(\cos\zeta + \cos\xi)}{2} & \dfrac{\sqrt{3}(\cos\xi - \cos\zeta)}{2} \\ \dfrac{\sqrt{3}(\cos\xi - \cos\zeta)}{2} & 3 - \dfrac{3(\cos\zeta + \cos\xi)}{2} \end{pmatrix}.$$

where

$$\zeta = \frac{k_1 l}{2} + \frac{\sqrt{3}}{2} k_2 l \quad \text{and} \quad \xi = \frac{k_1 l}{2} - \frac{\sqrt{3}}{2} k_2 l. \tag{6.52}$$

We note that the dispersion equation is bi-quadratic with respect to ω, and it has the form

$$\omega^4(m^2 - \alpha^2) - \omega^2 m \,\mathrm{tr}\mathbf{C} + \det\mathbf{C} = 0. \tag{6.53}$$

Since $c > 0$, both the trace $\mathrm{tr}\mathbf{C}$ and the determinant $\det\mathbf{C}$ are positive for all k_1 and k_2 in the elementary cell of the reciprocal lattice, except at the origin where they are both zero. Hence there are two important regimes. For $m^2 > \alpha^2$, there are two dispersion surfaces. As the factor $(m^2 - \alpha^2)$ decreases towards a critical point where $m^2 = \alpha^2$, the upper dispersion surface, and its gradient, increase, and the lower surface decreases. At this critical point, the gradient of the upper surface becomes infinite, and the equation for the lower surface becomes $\omega = (m^{-1} \det\mathbf{C}/\mathrm{tr}\mathbf{C})^{1/2}$. For $m^2 < \alpha^2$, only the lower dispersion surface remains. We refer to the regimes $m^2 > \alpha^2$ or $m^2 < \alpha^2$ as *subcritical* or *supercritical*, respectively.

We also note that in the supercritical regime, where only one dispersion surface remains, the dynamics of the active chiral lattice is dominated by the shear waves associated with the vortex-type waveforms. In this case, the lattice acts as a polariser supporting only shear waves.

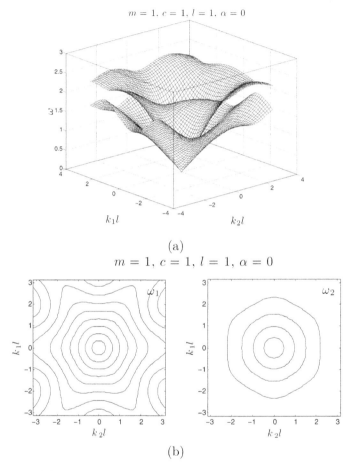

FIGURE 6.4: (a) Dispersion surfaces for the monatomic lattice with no spinners. (b) Slowness contours for the two eigenfrequencies for the monatomic lattice with no spinners.

6.2.4 Dispersion properties of the monatomic lattice

Firstly, Figure 6.4 gives dispersion surfaces and the corresponding two-dimensional diagrams with the slowness contours for the uniform lattice without spinners ($\alpha = 0$). Two conical surfaces are clearly visible in the low frequency range. These surfaces are shown to evolve as the spinners are brought into the system, now with chiral terms in the equations of motion.

6.2.4.1 Lattice of the vortex-type

The diagrams for the vortex-type of monatomic lattice, with a small spinner constant* $\alpha = 0.5$ and mass $m = 1$ are given in Figure 6.5. It is apparent that the lower surface, dominated by shear waves, does not change by much as a result of the vortex interaction. On the contrary, the upper acoustic surface, dominated by pressure waves extends further into a higher range of frequencies and has a pronounced conical shape for a vortex-type lattice of small α. The latter implies that the homogenisation range for Bloch–Floquet waves corresponding to this dispersion surface is also extended into a wider range of frequencies, as shown in Figure 6.5.

For the other regime, when $m^2 < \alpha^2$, the dispersion diagram illustrating such a situation for the case of $\alpha = 2$ and $m = 1$ is given in Figure 6.6. Only a single dispersion surface is present on this diagram.

The influence of the vortex-type interaction appears to be significant for the dispersion properties of Bloch–Floquet waves within the entire admissible frequency range. The illustration is given by Figure 6.7 where we assume $k_1 = k_2 = k$ and plot ω as a function of the spinner constant α and the Bloch parameter k. In this case, the lower dispersion surface is represented by the diagram in Figure 6.7a and is defined for all real values of α whereas the upper dispersion surface disappears in the supercritical regime when $|\alpha| \geq m$. The diagram characterising the upper dispersion surface and showing ω as a function of α and k is presented in Figure 6.7b for $0 \leq \alpha < m$ where $m = 1$. It is noted that for a fixed value of k, the frequency ω on the upper dispersion surface increases with the increase of the spinner constant α, as shown in Figure 6.7b, whereas Figure 6.7a demonstrates that on the lower dispersion surface the frequency ω decreases as the spinner constant α increases.

6.2.4.2 Low-frequency range

We examine the behaviour of $\omega(\mathbf{k})$ in the low frequency limit for small values of $|\mathbf{k}|$, and evaluate the effective group velocity, corresponding to a quasi-static response of a homogenised elastic solid. The solutions to Equation (6.53) may be expanded for small values of k_1 and k_2 as follows

---

*Normalisation has been used throughout this section, so that $c = 1$, $m = 1$ and the distance between the neighbouring masses is equal to unity. All other physical quantities have been normalised accordingly, and the physical units are not shown.

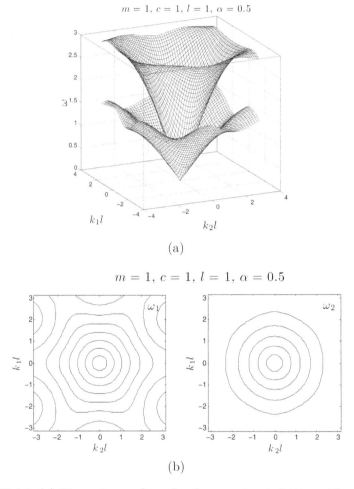

FIGURE 6.5: (a) Dispersion surfaces for the monatomic lattice with spinners $\alpha = 0.5$. (b) Slowness contours for the two eigenfrequencies for the monatomic lattice with spinners $\alpha = 0.5$.

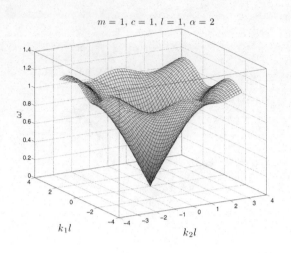

FIGURE 6.6: Dispersion surface for the monatomic lattice with spinners $\alpha = 2$.

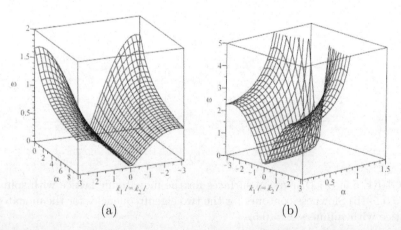

FIGURE 6.7: The radian frequency change at the lower dispersion surface (a) and the upper dispersion surface (b) with α along the line $k_1 = k_2$ in the reciprocal space. $m = 1$, $c = 1$ and $l = 1$.

$$\omega_1 = \sqrt{\frac{3c}{8}\left(\frac{2m-(m^2+3\alpha^2)^{1/2}}{(m^2-\alpha^2)}\right)((k_1l)^2+(k_2l)^2)}, \qquad \alpha \neq m, \qquad (6.54)$$

$$\omega_1 = \frac{3}{8}\sqrt{\frac{2c}{m}((k_1l)^2+(k_2l)^2)}, \qquad \alpha = m, \qquad (6.55)$$

and

$$\omega_2 = \sqrt{\frac{3c}{8}\left(\frac{2m+(m^2+3\alpha^2)^{1/2}}{(m^2-\alpha^2)}\right)((k_1l)^2+(k_2l)^2)}, \qquad \alpha < m.$$

For the case when no spinners are present, these formulae reduce to

$$\omega_1^0 = \sqrt{\frac{3c}{8m}((k_1l)^2+(k_2l)^2)} \quad \text{and} \quad \omega_2^0 = \sqrt{3}\,\omega_1^0.$$

These expressions for ω_1^0 and ω_2^0 are consistent with the special case of Equation (3.3) in [49].

In the supercritical regime $m^2 \leq \alpha^2$, for low frequencies, only one dispersion surface exists representing a shear wave. For a monatomic, harmonic scalar lattice (with no spinners), the dispersion equation is given in [8] as

$$\omega = \sqrt{\frac{c_s}{m_s}\left(8 - 4\cos^2\frac{k_1l}{2} - 4\cos\frac{k_1l}{2}\cos\frac{\sqrt{3}k_2l}{2}\right)},$$

where the subscript s corresponds to the scalar case of out-of-plane shear. The low frequency approximation for this dispersion equation is

$$\omega = \sqrt{\frac{3c_s}{2m_s}((k_1l)^2+(k_2l)^2)}. \qquad (6.56)$$

It is interesting to compare Equations (6.54)–(6.56) in the regime $m^2 \leq \alpha^2$. Formally, the system under consideration is a vector system corresponding to an elastic lattice. However, in the regime when $m^2 \leq \alpha^2$, there exists only one dispersion surface. Therefore, we may compare this system with an "equivalent" scalar lattice in which the waves have the same dispersion properties as the shear waves in the vector lattice in the low frequency regime. Comparing Equations (6.54) and (6.56) and by the appropriate choice

$$\frac{c_s}{m_s} = \frac{c}{4m}\frac{\sqrt{1+3(\alpha/m)^2}-2}{(\alpha/m)^2-1}$$

of combination of parameters c_s, c and m_s, m for both lattices together with α for the spinners in the vector lattice, the shear wave in the vector lattice and the wave in the scalar lattice may be constructed so as to have the same dispersive properties.

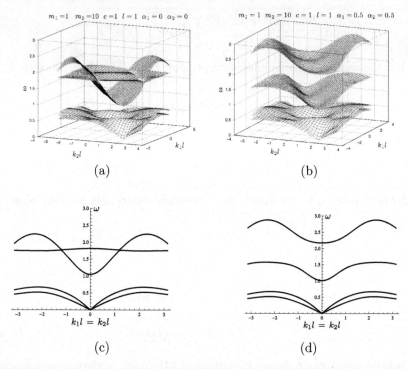

FIGURE 6.8: Dispersion surfaces for the biatomic lattice (a) with contrasting masses and no spinners, and (b) with contrasting masses and the same spinner throughout the lattice. (c) and (d) are the cross-section of the surfaces given in (a) and (b), respectively, for $k_1 l = k_2 l$.

6.2.4.3 Biatomic lattice of the vortex-type

Now we show the computations for Bloch–Floquet waves for the biatomic triangular lattice described in Section 6.2.1. The dispersion equation is given by relation (6.50), where

$$\mathbf{M} = \mathrm{diag}[m_1, m_1, m_2, m_2]$$

and

$$\Sigma(\alpha) = \begin{pmatrix} 0 & -i\alpha_1 & 0 & 0 \\ i\alpha_1 & 0 & 0 & 0 \\ 0 & 0 & 0 & -i\alpha_2 \\ 0 & 0 & i\alpha_2 & 0 \end{pmatrix}.$$

Additionally, the stiffness matrix can be expressed in the form (see [108])

$$\mathbf{C}(\mathbf{k}) = c \begin{pmatrix} \mathbf{C}_{11}(\mathbf{k}) & \mathbf{C}_{12}(\mathbf{k}) \\ \mathbf{C}_{21}(\mathbf{k}) & \mathbf{C}_{22}(\mathbf{k}) \end{pmatrix},$$

with the 2×2 matrices

$$\mathbf{C}_{11}(\mathbf{k}) = \mathbf{C}_{22}(\mathbf{k}) = \begin{pmatrix} 3 - \frac{1}{2}\cos\zeta & -\frac{\sqrt{3}}{2}\cos\zeta \\ -\frac{\sqrt{3}}{2}\cos\zeta & 3 - \frac{3}{2}\cos\zeta \end{pmatrix},$$

$$\mathbf{C}_{21}(\mathbf{k}) = \overline{\mathbf{C}_{12}(\mathbf{k})} = \exp[i(\zeta + \xi)] \begin{pmatrix} -2\cos(\zeta + \xi) - \frac{1}{2}\cos\xi & \frac{\sqrt{3}}{2}\cos\xi \\ \frac{\sqrt{3}}{2}\cos\xi & -\frac{3}{2}\cos\xi \end{pmatrix},$$

where ζ and ξ are given in Equation (6.52).

For such a lattice with no spinners, the dispersion surfaces for Bloch–Floquet waves are shown in Fig 6.8a (a section of the surfaces for $k_1 = k_2$ is added in Figure 6.8c). In contrast, for the vortex-type of lattice the corresponding results are given in Figure 6.8b (with a section for $k_1 = k_2$ in Figure 6.8d).

These computations show that the vortex-type interaction within the lattice leads to formation of additional total band gaps, as shown in Figure 6.8b. It is noted that the change in the behaviour of the two lowest dispersion surfaces through the introduction of spinners is small, whereas additional band gaps are introduced in the upper two dispersion surfaces.

The influence of the spinner constant α on the dispersive properties of Bloch–Floquet waves is illustrated in Figure 6.9 where ω is represented as a function of α and $kl = k_1 l = k_2 l$. The two lower shear dispersion surfaces, represented in Figure 6.9a, are defined for all real values of α, whereas the two upper pressure dispersion surfaces, depicted in Figure 6.9b, disappear when critical regimes determined by the spinner constant α are reached. In particular, in the *subcritical regime* ($\alpha \leq 1$) two pressure waves propagate while one disappears in the *intercritical regime* ($1 < \alpha \leq 10$) and no pressure waves propagate in the *supercritical regime* ($\alpha > 10$). Also, for the lower dispersion surfaces, at sufficiently large α the frequency ω decreases as the spinner constant α increases (see Figure 6.9a).

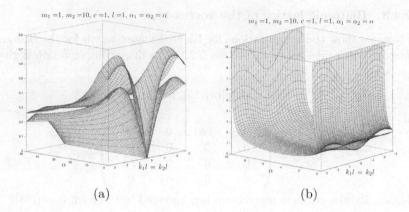

FIGURE 6.9: The radian frequency changes at the two lower dispersion surfaces (a) and the two upper dispersion surfaces (b) with α along the line $k_1 = k_2$ in the reciprocal space. $m_1 = 1$, $m_2 = 10$, $c = 1$ and $l = 1$.

6.3 Discrete structured interface: shielding, negative refraction, and focusing

In this section, applications of the dispersive properties of Bloch–Floquet waves in discrete systems are considered. In particular, applications relating to the effects of filtration and focussing of elastic waves by a *"metamaterial flat lens"* for certain frequencies are presented. The effects of focussing and filtering for solutions of the Helmholtz equation have already been demonstrated in the literature, see for example [110]. The paper [83] discussed similar effects for the case of vector elasticity in a structured continuum. Here, we follow the papers [49,50] and discuss the effects of focussing and filtration of elastic waves in discrete structures.

6.3.1 Equations of motion

We consider a planar triangular biatomic lattice where the elementary cell consists of two masses, and the masses are connected by identical Euler–Bernoulli beams. With a little generalisation, the governing equations can be formulated in the same manner as in Section 4.2.2. In particular, we now label each mass by the multi-index $\boldsymbol{m} \in \mathbb{Z}^2$ and a scalar $n \in \{1, 2\}$. The multi-index $\boldsymbol{m}$ refers to the unit cell in which the particle is located, whereas the scalar n distinguishes between different particles in the same unit cell. The position of each particle in the lattice is then $\boldsymbol{x}(\boldsymbol{m}, n) = \mathsf{T}\boldsymbol{m} + \boldsymbol{x}_0(n)$, where $\mathsf{T} = (\boldsymbol{t}_1, \boldsymbol{t}_2)$ and $\boldsymbol{x}_0(n) = (\delta_{n,2}, 0)^\mathsf{T}$ is the location of particle n in cell $\boldsymbol{m} = \boldsymbol{0}$, and $\boldsymbol{t}_1 = (2\ell, 0)^\mathsf{T}$ and $\boldsymbol{t}_2 = (\ell/2, \ell\sqrt{3}/2)^\mathsf{T}$ are the direct lattice vectors. The

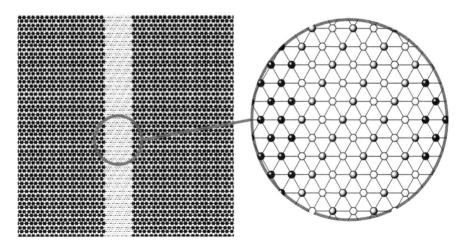

FIGURE 6.10: A schematic diagram of the lattice system with the heterogeneous biatomic interface (highlighted). The regions to the left and to the right of the interface consist of homogeneous monatomic lattices.

governing equations for the time-harmonic motion of particle (m, n) have the form
$$-\omega^2 \mathsf{M}(n)\boldsymbol{u}_{m,n} + \sum_{(\boldsymbol{p},q)\in\mathcal{N}(n)} \mathsf{A}(\boldsymbol{p},q)\boldsymbol{u}_{\boldsymbol{m}+\boldsymbol{p},q} = \mathbf{0}, \qquad (6.57)$$

where $\mathcal{N}(n)$ is the set of particles $(\boldsymbol{m}+\boldsymbol{p}, q)$ connected to node $(\boldsymbol{m}, n)$ and the symbols have the same meaning as in Section 4.2.2. Applying the discrete Fourier transform (cf. 4.27) to (6.57) yields

$$\sum_{(\boldsymbol{p},q)\in\mathcal{N}(n)} \left[\mathsf{A}(\boldsymbol{p},q)e^{-i\boldsymbol{k}\cdot\mathsf{T}\boldsymbol{p}} - \omega^2 \mathsf{M}(n)\delta_{n,q}\right] \boldsymbol{u}_n^{\mathrm{F}}(\boldsymbol{k}) = \mathbf{0}, \qquad (6.58)$$

whence the dispersion equation for a perfect lattice is immediately obtained as
$$\det\left(\sigma(\boldsymbol{k}) - \omega^2 \mathcal{M}\right) = 0, \qquad (6.59)$$

where the matrix σ is partitioned as

$$\sigma_{nq} = \sum_{\boldsymbol{p}\in\mathcal{N}(n)} \left[\mathsf{A}(\boldsymbol{p},q)e^{-i\boldsymbol{k}\cdot\mathsf{T}\boldsymbol{p}}\right] \quad \text{and} \quad \mathcal{M}_{nq} = \mathsf{M}(n)\delta_{n,q}.$$

For the case when the lattice is forced, the equation of motion (6.57) becomes

$$-\omega^2 \mathsf{M}(n)\boldsymbol{g}_{m,n} + \sum_{(\boldsymbol{p},q)\in\mathcal{N}(n)} \mathsf{A}(\boldsymbol{p},q)\boldsymbol{g}_{\boldsymbol{m}+\boldsymbol{p},q} = \boldsymbol{F}_n \delta_{m,0}, \qquad (6.60)$$

where $\boldsymbol{F}_n$ is the force applied to node $(\mathbf{0}, n)$ and $\boldsymbol{g}_{m,n}$ is the displacement.

TABLE 6.1: The material and geometrical parameters for the ambient and interface lattices.

Property	Value
Young's Modulus	200 GPa
Second Moment of Inertia	349×10^{-8} m^4
Cross Sectional Area	2.12×10^{-3} m^2
Beam Density	7850 kg m^{-2}
Beam Length	1 m
Nodal Mass (Ambient)	91.531 kg
Nodal Mass (Interface m_1)	16.642 kg
Nodal Mass (Interface m_2)	166.42 kg
Polar Mass Moment of Inertia (Ambient)	66.568 kg m^2
Polar Mass Moment of Inertia (Interface J_1)	633.284 kg m^2
Polar Mass Moment of Inertia (Interface J_2)	99.852 kg m^2

Note: The parameters of the ambient and interface nodes and links are differentiated where required and are uniform otherwise.

Following the same procedure as for the homogeneous problem (6.57), we obtain the Green's tensor

$$g_{m,n} = \int_{-\pi/(2\ell)}^{\pi/(2\ell)} dk_1 \int_{-\pi/(\ell\sqrt{3})}^{\pi/(\ell\sqrt{3})} dk_2 \left[\sigma(\boldsymbol{k}) - \omega^2 \mathcal{M}\right]^{-1} \boldsymbol{F}_n. \qquad (6.61)$$

6.3.2 Constructing the structured interface

Consider a triangular biatomic lattice, as shown in Figure 6.10. Let the ambient lattice be monatomic and homogeneous. Within the ambient lattice a finite slab of heterogeneous biatomic lattice of the same geometry is embedded. Both the ambient lattice and interface lattice (finite slab) are lattices with inertial links, formed from Euler–Bernoulli beams. The material and geometrical parameters of the lattices are detailed in Table 6.1. A schematic diagram of the ambient and interface lattices is shown in Figure 6.10.

Consider the time-harmonic propagation of elastic in-plane waves through the ambient lattice and structural interface as shown in Figure 6.11b. The wave is generated by a single point source: a time-harmonic displacement of amplitude 10^{-6} m in the horizontal direction is prescribed at one of the lattice nodes. Damping is applied to the lattice links in the neighbourhood of the fixed boundary nodes in order to reduce reflection from the boundary of the computational domain.

Consider now the dispersion surfaces for the elementary cell of the structured interface, shown in Figure 6.12. The transmission problem is formally distinct from the Bloch–Floquet spectral problem. Nevertheless, the disper-

Structured interfaces and chiral systems in dynamics of elastic solids 223

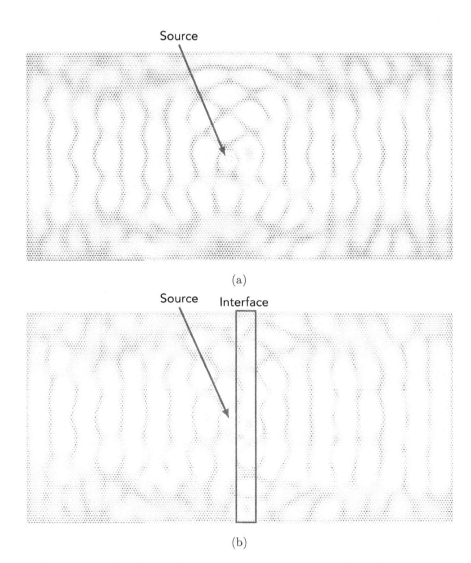

FIGURE 6.11: (a) shows a harmonic wave propagating through the ambient lattice. (b) shows a harmonic wave interacting with the structured interface. The magnitude of the displacement field is plotted. It is observed that the displacement field is essentially unaffected by the presence of the interface. The forcing frequency is 100 Hz.

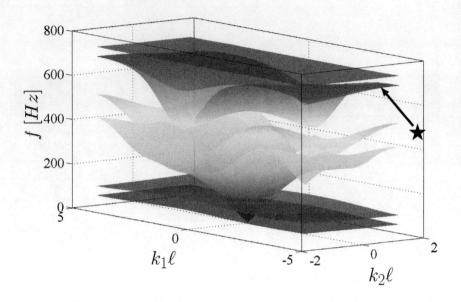

FIGURE 6.12: The dispersion surfaces corresponding to heterogeneous diatomic interface lattice. Of particular interest, in addition to the band gap at 700 Hz, is the surface labelled ★, which possesses saddle points.

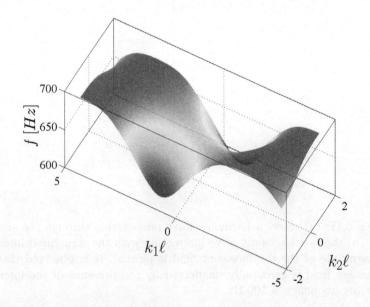

FIGURE 6.13: The sixth dispersion surface, labelled ★ in Figure 6.12, possessing the saddle point of interest.

Structured interfaces and chiral systems in dynamics of elastic solids 225

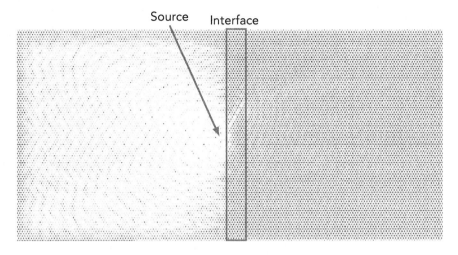

FIGURE 6.14: The same configuration as in Figure 6.11b, but with a forcing frequency of 700 Hz, which lies in the band gap of the dispersion diagram for the interface lattice. The wave is reflected from the interface as would be expected for frequencies within the stop band for the interface.

sion diagram may be used in order to predict the reflection and transmission patterns. Figure 6.11b shows the modulus of the displacement vector $|u(x)|$ when the forcing frequency is 100 Hz. A similar wave pattern can clearly be observed on both sides of the interface layer, indicating that the low frequency response of the structured interface is very close to that of the ambient medium. A similar wave pattern can also be seen in Figure 6.11a where the structured interface has been removed entirely. In contrast, Figure 6.14 shows the magnitude of the displacement field when the forcing frequency is 700 Hz, which lies in the stop band of Figure 6.12. In this case, the incoming wave is reflected, with very little transmission.

As discussed in [49,50] and [49,83,110], the phenomenon of focussing by a flat interface is linked to the presence of saddle points and regions of negative group velocities. Referring to the dispersion surfaces for the heterogeneous interface lattice, as in Figure 6.12, it is observed that the surface labelled ★ and shown in Figure 6.13 possesses a saddle point and regions where the group velocity is negative. In particular, for small perturbations around the saddle point it is observed that the components of the group velocity $(\partial \omega / \partial k)$ will have opposing signs. Figure 6.15 shows a plot of the magnitude of the displacement field when the forcing frequency is 642.5 Hz. The frequency was chosen in the vicinity of the saddle point on the corresponding dispersion surface. The effect described here is typical for neighbourhoods of saddle points. A clear directional preference can be observed within the interface. In addition, the *secondary source* on the right hand side of the interface can also be observed.

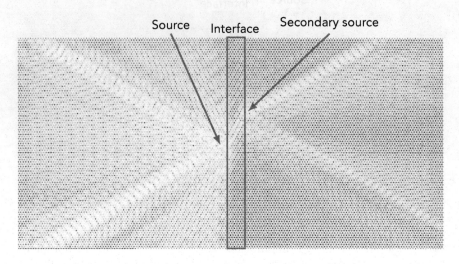

FIGURE 6.15: The same configuration as in Figure 6.11b, but with a forcing frequency of 642.5 Hz. The secondary source is visible on the right-hand side of the interface and is shifted along the direction of preferential propagation of the interface lattice.

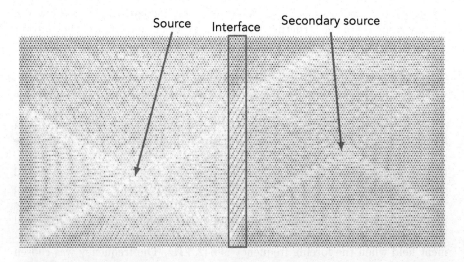

FIGURE 6.16: In this case the source has been shifted further away from the interface. Correspondingly, this leads to a shift in the image point due to the preferential direction within the layer. Here, the forcing frequency is 654.5 Hz.

Figure 6.15 shows the preferential direction of propagation and the effect of focussing. This feature of the waves persists in a small interval containing the saddle point.

Finally, in Figure 6.16 a simulation where the source has been shifted away from the interface region is presented. In this case, the forcing frequency is 654.4 Hz, which again is in the vicinity of the saddle point and within the region where there is a preferential direction of propagation. Moreover, where the beams intersect on the right hand side, the formation of the *image point* becomes apparent.

We hope the reader has enjoyed the book and found the material both stimulating and interesting. In particular, for research students we trust that this book has brought new ideas, which can be used for novel projects. The research area of modelling waves in multi-scale media is very rich and, by no means, does the material of this book cover it all. Additionally, the range of possible practical applications is immense, and we hope that physicists, engineers and industrial scientists will find this book to be of great value.

References

[1] R Abdelmoula and J Marigo. The effective behavior of a fiber bridged crack. *Journal of the Mechanics and Physics of Solids*, 48(11):2419–2444, 2000.

[2] I D Abrahams and G R Wickham. The propagation of elastic waves in a certain class of inhomogeneous anisotropic materials. i. the refraction of a horizontally polarized shear wave source. *Proceedings of the Royal Society of London A: Mathematical, Physical and Engineering Sciences*, 436(1898):449–478, 1992.

[3] D J Acheson. *Elementary Fluid Dynamics*. Oxford University Press, Oxford, UK, 1990.

[4] Y A Antipov, O Avila-Pozos, S T Kolaczkowski, and A B Movchan. Mathematical model of delamination cracks on imperfect interfaces. *International Journal of Solids and Structures*, 38(36–37):6665–6697, 2001.

[5] T Antonakakis and R V Craster. High-frequency asymptotics for microstructured thin elastic plates and platonics. *Proceedings of the Royal Society of London A: Mathematical, Physical and Engineering Sciences*, 468(2141):1408–1427, 2012.

[6] T Antonakakis, R V Craster, and S R L Guenneau. Homogenisation for elastic photonic crystals and dynamic anisotropy. *Journal of the Mechanics and Physics of Solids*, 71:84–96, 2014.

[7] O Avila-Pozos, A Klarbring, and A B Movchan. Asymptotic model of orthotropic highly inhomogeneous layered structure. *Mechanics of Materials*, 31(2):101–116, 1999.

[8] M V Ayzenberg-Stepanenko and L I Slepyan. Resonant-frequency primitive waveforms and star waves in lattices. *Journal of Sound and Vibration*, 313(3):812–821, 2008.

[9] M D Bacon, P Dean, and J L Martin. Defect modes in two-component chains. *Proceedings of the Royal Society of London A: Mathematical, Physical and Engineering Sciences*, 80:174–189, 1962.

[10] N Bakhvalov and G Panasenko. *Homogenization: Averaging Processes in Periodic Media*. Kluwer, Dordrecht, 1989.

[11] C M Bender and S A Orszag. *Advanced Mathematical Methods for Scientists and Engineers*. McGraw-Hill, New York, 1978.

[12] A Bensoussan, J Lions, and G Papanicolaou. *Asymptotic Analysis for Periodic Structures*, volume 374. American Mathematical Society, 2011.

[13] Y Benveniste and T Chen. On the Saint-Venant torsion of composite bars with imperfect interfaces. *Proceedings of the Royal Society of London A: Mathematical, Physical and Engineering Sciences*, 457(2005):231–255, 2001.

[14] Y Benveniste and T Miloh. Neutral inhomogeneities in conduction phenomena. *Journal of the Mechanics and Physics of Solids*, 47(9):1873–1892, 1999.

[15] Y Benveniste and T Miloh. Imperfect soft and stiff interfaces in two-dimensional elasticity. *Mechanics of Materials*, 33(6):309–323, 2001.

[16] K Bertoldi, D Bigoni, and W J Drugan. A discrete-fibers model for bridged cracks and reinforced elliptical voids. *Journal of the Mechanics and Physics of Solids*, 55(5):1016–1035, 2007.

[17] K Bertoldi, D Bigoni, and W J Drugan. Structural interfaces in linear elasticity. Part I: Nonlocality and gradient approximations. *Journal of the Mechanics and Physics of Solids*, 55(1):1–34, 2007.

[18] K Bertoldi, D Bigoni, and W J Drugan. Structural interfaces in linear elasticity. Part II: effective properties and neutrality. *Journal of the Mechanics and Physics of Solids*, 55(1):35–63, 2007.

[19] D Bigoni and M Gei. Bifurcations of a coated, elastic cylinder. *International Journal of Solids and Structures*, 38(30-31):5117–5148, 2001.

[20] D Bigoni, M Gei, and A B Movchan. Dynamics of a prestressed stiff layer on an elastic half space: filtering and band gap characteristics of periodic structural models derived from long-wave asymptotics. *Journal of the Mechanics and Physics of Solids*, 56(7):2494–2520, 2008.

[21] D Bigoni and A B Movchan. Statics and dynamics of structural interfaces in elasticity. *International Journal of Solids and Structures*, 39(19):4843–4865, 2002.

[22] D Bigoni, M Ortiz, and A Needleman. Effect of interfacial compliance on bifurcation of a layer bonded to a substrate. *International Journal of Solids and Structures*, 34(33):4305–4326, 1997.

[23] D Bigoni, S K Serkov, M Valentini, and A B Movchan. Asymptotic models of dilute composites with imperfectly bonded inclusions. *International Journal of Solids and Structures*, 35(24):3239–3258, 1998.

[24] J Billingham and A C King. *Wave Motion*, volume 24. Cambridge University Press, Cambridge, 2001.

[25] M Born and E Wolf. *Principles of Optics*. Pergamon, London, 1st edition, 1959.

[26] L Brillouin. *Wave Propagation in Periodic Structures*. Dover, New York, 2nd edition, 1953.

[27] M Brun, D J Colquitt, I S Jones, A B Movchan, and N V Movchan. Transformation cloaking and radial approximations for flexural waves in elastic plates. *New Journal of Physics*, 16(9):093020, 2014.

[28] M Brun, S R L Guenneau, and A B Movchan. Achieving control of in-plane elastic waves. *Applied Physics Letters*, 94(6):061903–061903, 2009.

[29] M Brun, S R L Guenneau, A B Movchan, and D Bigoni. Dynamics of structural interfaces: Filtering and focussing effects for elastic waves. *Journal of the Mechanics and Physics of Solids*, 58(9):1212–1224, 2010.

[30] M Brun, I S Jones, and A B Movchan. Vortex-type elastic structured media and dynamic shielding. *Proceedings of the Royal Society of London A: Mathematical, Physical and Engineering Sciences*, 468:20120165, 2012.

[31] M Brun, A B Movchan, and N V Movchan. Shear polarisation of elastic waves by a structured interface. *Continuum Mechanics and Thermodynamics*, 22(6–8):663–677, 2010.

[32] B Budiansky and G F Carrier. The pointless wedge. *SIAM Journal on Applied Mathematics*, 25(3):378–387, 1973.

[33] W Bühring. Generalized hypergeometric functions at unit argument. *Proceedings of the American Mathematical Society*, 114(1):145–153, 1992.

[34] S Caddemi and Calio I. Exact solution of the multi-cracked euler-bernoulli column. *International Journal of Solids and Structures*, 45(5):1332–1351, 2008.

[35] J Callaway. Theory of scattering in solids. *Journal of Mathematical Physics*, 5(6):783–798, 1964.

[36] J Callaway. *Quantum Theory of the Solid State*. Academic Press, New York, 1974.

[37] G A Campbell. Physical theory of the electric wave-filter. *Bell System Technical Journal*, 1(2):1–32, 1922.

[38] A Cantoni and P Butler. Eigenvalues and eigenvectors of symmetric centrosymmetric matrices. *Linear Algebra and Its Applications*, 13:275–288, 1976.

[39] G Carta, M Brun, A B Movchan, N V Movchan, and I S Jones. Dispersion properties of vortex-type monatomic lattices. *International Journal of Solids and Structures*, 51(11):2213–2225, 2014.

[40] G Carta, I S Jones, N V Movchan, A B Movchan, and M J Nieves. "Deflecting elastic prism" and unidirectional localisation for waves in chiral elastic systems. *Scientific Reports*, 7(1):26, 2017.

[41] J Casado-Diaz, M Luna-Laynez, and F Murat. Asymptotic behavior of an elastic beam fixed on a small part of one of its extremities. *Comptes Rendus Mathematique*, 338(12):975–980, 2004.

[42] C T Chan, F Huang, Xand Liu, and Z H Hang. Dirac-dispersion and zero-index in two dimensional and three dimensional photonic and phononic systems. *Progress In Electromagnetics Research B*, 44:163–190, 2012.

[43] H Chen, C T Chan, and P Sheng. Transformation optics and metamaterials. *Nature Materials*, 9(5):387–396, 2010.

[44] A Colombi, D J Colquitt, P Roux, S R L Guenneau, and R V Craster. A seismic metamaterial: The resonant metawedge. *Scientific Reports*, 6:27717, 2016.

[45] D J Colquitt, M Brun, M Gei, A B Movchan, N V Movchan, and I S Jones. Transformation elastodynamics and cloaking for flexural waves. *Journal of the Mechanics and Physics of Solids*, 72:131–143, 2014.

[46] D J Colquitt, A Colombi, R V Craster, P Roux, and S R L Guenneau. Seismic metasurfaces: Sub-wavelength resonators and Rayleigh wave interaction. *Journal of the Mechanics and Physics of Solids*, 99:379–393, 2017.

[47] D J Colquitt, R V Craster, and M Makwana. High frequency homogenisation for elastic lattices. *The Quarterly Journal of Mechanics and Applied Mathematics*, 68(2):203–230, 2015.

[48] D J Colquitt, I S Jones, N V Movchan, A B Brun, and R C McPhedran. Making waves round a structured cloak: lattices, negative refraction and fringes. *Proceedings of the Royal Society A: Mathematical, Physical and Engineering Science*, 469:20130218, 2013.

[49] D J Colquitt, I S Jones, N V Movchan, and A B Movchan. Dispersion and localization of elastic waves in materials with microstructure. *Proceedings of the Royal Society A: Mathematical, Physical and Engineering Science*, 467(2134):2874–2895, October 2011.

[50] D J Colquitt, I S Jones, N V Movchan, A B Movchan, and R C McPhedran. Dynamic anisotropy and localization in elastic lattice systems. *Waves in Random and Complex Media*, 22(2):143–159, 2012.

[51] D J Colquitt, A B Movchan, I S Jones, and W Daniels. Frequency related localisation of harmonic elastic waves in stratified welds. *Acta Mechanica Sinica*, 26(4):567–572, 2010.

[52] D J Colquitt, N V Movchan, and A B Movchan. Parabolic metamaterials and dirac bridges. *Journal of the Mechanics and Physics of Solids*, 95:621–631, 2016.

[53] D J Colquitt, M J Nieves, I S Jones, A B Movchan, and N V Movchan. Localization for a line defect in an infinite square lattice. *Proceedings of the Royal Society of London A: Mathematical, Physical and Engineering Sciences*, 469(2150):20120579, 2013.

[54] R V Craster, S R L Guenneau, J Kaplunov, and E Nolde. On a class of three-phase checkerboards with unusual effective properties. *Comptes Rendus Mécanique*, 339(6):411–417, 2011.

[55] R V Craster, J Kaplunov, and A V Pichugin. High-frequency homogenization for periodic media. *Proceedings of the Royal Society A: Mathematical, Physical and Engineering Science*, 466(2120):2341–2362, 2010.

[56] R V Craster, J Kaplunov, and J Postnova. High-frequency asymptotics, homogenisation and localisation for lattices. *The Quarterly Journal of Mechanics and Applied Mathematics*, 63(4):497–519, 2010.

[57] S A Cummer and D Schurig. One path to acoustic cloaking. *New Journal of Physics*, 9(3):45–45, 2007.

[58] L S Dolin. To the possibility of comparison of three-dimensional electromagnetic systems with nonuniform anisotropic filling. *Izvestiya Vysshikh Uchebnykh Zavedeni Radiofizika*, 4(5):964–967, 1961.

[59] K B Dossou, L C Botten, R C McPhedran, and C G Poulton. Shallow defect states in two-dimensional photonic crystals. *Physical Review A*, 77:1–18, 2008.

[60] J P Dowling. Photonic & sonic band-gap and metamaterial bibliography. http://phys.lsu.edu/~jdowling/pbgbib.html.

[61] D V Evans and R Porter. Penetration of flexural waves through a periodically constrained thin elastic plate in vacuo and floating on water. *Journal of Engineering Mathematics*, 58(1-4):317–337, 2007.

[62] N Fang, H Lee, C Sun, and X Zhang. Sub–diffraction-limited optical imaging with a silver superlens. *Science*, 308(5721):534–537, 2005.

[63] M Farhat, S Enoch, S R L Guenneau, and A B Movchan. Broadband cylindrical acoustic cloak for linear surface waves in a fluid. *Physical Review Letters*, 101(13):134501, 2008.

[64] M Farhat, S R L Guenneau, and S Enoch. Ultrabroadband elastic cloaking in thin plates. *Physical Review Letters*, 103(2):24301, 2009.

[65] M Farhat, S R L Guenneau, S Enoch, and A B Movchan. Cloaking bending waves propagating in thin elastic plates. *Physical Review B*, 79(3):033102, 2009.

[66] M Farhat, S R L Guenneau, S Enoch, A B Movchan, and G G Pétursson. Focussing bending waves via negative refraction in perforated thin plates. *Applied Physics Letters*, 96(8), 2010.

[67] M Farhat, S R L Guenneau, S Enoch, A B Movchan, F Zolla, and A Nicolet. A homogenization route towards square cylindrical acoustic cloaks. *New Journal of Physics*, 10(11):115030, 2008.

[68] M Gei, A B Movchan, and D Bigoni. Band-gap shift and defect-induced annihilation in prestressed elastic structures. *Journal of Applied Physics*, 105(6):063507, 2009.

[69] M Gei, A B Movchan, and I S Jones. Junction conditions for cracked elastic thin solids under bending and shear. *The Quarterly Journal of Mechanics and Applied Mathematics*, 62(4):481–493, 2009.

[70] G Geymonat and F Krasucki. Analyse asymptotique du comportement en flexion de deux plaques collées. *Comptes Rendus de l'Academie des Sciences - Series IIB - Mechanics-Physics-Chemistry-Astronomy*, 325(6):307–314, 1997.

[71] G Geymonat, F Krasucki, and S Lenci. Mathematical analysis of a bonded joint with a soft thin adhesive. *Mathematics and Mechanics of Solids*, 4(2):201–225, 1999.

[72] H Goldstein, C P Poole, and J L Safko. *Classical Mechanics*. Addison Wesley, Boston, 3rd edition, 2000.

[73] J W Goodman. *Introduction to Fourier Optics*. Roberts & Company, Colorado, 3rd edition, 2005.

[74] A Grbic and G V Eleftheriades. Subwavelength focusing using a negative-refractive-index transmission line lens. *Antennas and Wireless Propagation Letters, IEEE*, 2(1):186–189, 2003.

[75] A Greenleaf, M Lassas, and G Uhlmann. On nonuniqueness for Calderon's inverse problem. *Mathematical Research Letters*, 10:685–694, 2003.

[76] S R L Guenneau, A B Movchan, G G Pétursson, and S A Ramakrishna. Acoustic metamaterials for sound focusing and confinement. *New Journal of Physics*, 9(11):399, 2007.

[77] S R L Guenneau, A B Movchan, F Zolla, N V Movchan, and A Nicolet. Acoustic band gaps in arrays of neutral inclusions. *Journal of Computational and Applied Mathematics*, 234(6):1962–1969, 2010.

[78] Z Hashin. Thermoelastic properties of fiber composites with imperfect interface. *Mechanics of Materials*, 8(4):333–348, 1990.

[79] Z Hashin. Extremum principles for elastic heterogenous media with imperfect interfaces and their application to bounding of effective moduli. *Journal of the Mechanics and Physics of Solids*, 40(4):767–781, 1992.

[80] V Jacques, E Wu, F Grosshans, F Treussart, P Grangier, A Aspect, and J F Roch. Experimental realization of Wheeler's delayed-choice gedanken experiment. *Science*, 315(5814):966–968, 2007.

[81] I S Jones, M Brun, N V Movchan, and A B Movchan. Singular perturbations and cloaking illusions for elastic waves in membranes and Kirchhoff plates. *International Journal of Solids and Structures*, 69:498–506, 2015.

[82] I S Jones and A B Movchan. Bloch–floquet waves and controlled stop bands in periodic thermo-elastic structures. *Waves in Random and Complex Media*, 17(4):429–438, 2007.

[83] I S Jones, A B Movchan, and M Gei. Waves and damage in structured solids with multi-scale resonators. *Proceedings of the Royal Society A: Mathematical, Physical and Engineering Science*, 467(2128):964–984, 2011.

[84] G S Joyce and I J Zucker. Evaluation of the Watson integral and associated logarithmic integral for the d-dimensional hypercubic lattice. *Journal of Physics A: Mathematical and General*, 34:7349–7354, 2001.

[85] C Kittel. *Introduction to Solid State Physics*. John Wiley & Sons, London, 7th edition, 1995.

[86] A Klarbring. Derivation of a model of adhesively bonded joints by the asymptotic expansion method. *International Journal of Engineering Science*, 29(4):493–512, 1991.

[87] A Klarbring and A B Movchan. Asymptotic modelling of adhesive joints. *Mechanics of Materials*, 28(1):137–145, 1998.

[88] R V Kohn, H Shen, M S Vogelius, and M I Weinstein. Cloaking via change of variables in electric impedance tomography. *Inverse Problems*, 24(1):015016, 2008.

[89] Y K Konenkov. Diffraction of a flexural wave by a circular obstacle in a plate. *Akusticheskii Zhurnal*, 10(2):186–190, 1964.

[90] V A Kozlov, V G Maz'ya, and A B Movchan. *Asymptotic Analysis of Fields in Multi-Structures*. Oxford Science Publications, Oxford, 1999.

[91] I A Kunin. *Elastic Media with Micro-Structure*, volume 1. Springer, Berlin, 1982.

[92] R S Langley. The response of two-dimensional periodic structures to point harmonic forcing. *Journal of Sound and Vibration*, 197(4):447–469, 1996.

[93] R S Langley. The response of two-dimensional periodic structures to impulsive point loading. *Journal of Sound and Vibration*, 201(2):235–253, 1997.

[94] R S Langley, N S Bardell, and H M Ruivo. The response of two-dimensional periodic structures to harmonic point loading: a theoretical and experimental study of a beam grillage. *Journal of Sound and Vibration*, 207(4):521–535, 1997.

[95] A W Leissa. Vibration of plates. Technical Report NASA-SP-160, Scientific and Technical Information Division, NASA, Washington, DC, 1969.

[96] S G Lekhnitskii, S W Tsai, and T Cheron. *Anistropic Plates*. Gordon and Breach, New York, 1968.

[97] J Lekner. *Theory of Reflection of Electromagnetic and Particle Waves*. Springer, 1987.

[98] U Leonhardt. Optical conformal mapping. *Science*, 312(5781):1777–1780, 2006.

[99] S C S Lin and T J Huang. Tunable phononic crystals with anisotropic inclusions. *Physical Review B*, 83:174303, 2011.

[100] R Lipton. Variational methods, bounds, and size effects for composites with highly conducting interface. *Journal of the Mechanics and Physics of Solids*, 45(3):361–384, 1997.

[101] R Lipton and B Bernescu. Variational methods, size effects and extremal microgeometries for elastic composites with imperfect interface. *Mathematical Models and Methods in Applied Sciences*, 5(8):1139–1173, 1995.

[102] S Mahmoodian, R C McPhedran, C de Sterke, K B Dossou, C G Poulton, and L C Botten. Single and coupled degenerate defect modes in two-dimensional photonic crystal band gaps. *Physical Review A*, 79(1):1–12, 2009.

[103] B J Maling, D J Colquitt, and R V Craster. The homogenisation of Maxwell's equations with applications to photonic crystals and localised waveforms on gratings. *Wave Motion*, 69:35–48, 2017.

[104] A A Maradudin. Some effects of point defects on the vibrations of crystal lattices. *Reports on Progress in Physics*, 28:331–380, 1965.

[105] A A Maradudin, I M Lifshitz, A M Kosevich, W Cochran, and M J P Musgrave. *Lattice Dynamics*. Benjamin, 1969.

[106] A A Maradudin, E W Montroll, G H Weiss, and I P Ipatova. *Theory of Lattice Dynamics in the Harmonic Approximation*. Academic Press, London, 1971.

[107] P A Martin. Discrete scattering theory: Green's function for a square lattice. *Wave Motion*, 43(7):619–629, 2006.

[108] P G Martinsson and A B Movchan. Vibrations of lattice structures and phononic band gaps. *The Quarterly Journal of Mechanics and Applied Mathematics*, 56(1):45–64, 2003.

[109] V G Maz'ya, S A Nazarov, and B A Plamenevskij. *Asymptotic Theory of Elliptic Boundary Value Problems in Singular Perturbed Domains*. Birkhauser Basel, 2001.

[110] R C McPhedran, L C Botten, J McOrist, A A Asatryan, C M de Sterke, and N A Nicorovici. Density of states functions for photonic crystals. *Physical Review E*, 69:016609, 2004.

[111] R C McPhedran, A B Movchan, and N V Movchan. Platonic crystals: Bloch bands, neutrality and defects. *Mechanics of Materials*, 41(4):356–363, 2009.

[112] R C McPhedran, A B Movchan, N V Movchan, M Brun, and M J A Smith. 'Parabolic' trapped modes and steered Dirac cones in platonic crystals. *Proceedings of the Royal Society of London A: Mathematical, Physical and Engineering Sciences*, 471(2177):20140746, 2015.

[113] D J Mead. A general theory of harmonic wave propagation in linear periodic systems with multiple coupling. *Journal of Sound and Vibration*, 27:235–260, 1973.

[114] D J Mead. Wave propagation in continuous periodic structures: Research contributions from Southampton, 1964–1995. *Journal of Sound and Vibration*, 190:495–524, 1996.

[115] D J Mead and S Parthan. Free wave propagation in two-dimensional periodic plates. *Journal of Sound and Vibration*, 64:325–348, 1979.

[116] D J Mead and Y Yaman. The harmonic response of rectangular sandwich plates with multiply stiffening a flexural wave analysis. *Journal of Sound and Vibration*, 145:409–428, 1991.

[117] J Mei, Y Wu, C T Chan, and Z Zhang. First-principles study of Dirac and Dirac-like cones in phononic and photonic crystals. *Physical Review B*, 86(3):035141, 2012.

[118] D Melville and R Blaikie. Super-resolution imaging through a planar silver layer. *Optics Express*, 13(6):2127–2134, 2005.

[119] G W Milton, M Briane, and JR Willis. On cloaking for elasticity and physical equations with a transformation invariant form. *New Journal of Physics*, 8(10):248, 2006.

[120] G W Milton and N A Nicorovici. On the cloaking effects associated with anomalous localized resonance. *Proceedings of the Royal Society of London A: Mathematical, Physical and Engineering Sciences*, 462(2074):3027–3059, 2006.

[121] G W Milton, N A Nicorovici, R C McPhedran, K Cherednichenko, and Z Jacob. Solutions in folded geometries, and associated cloaking due to anomalous resonance. *New Journal of Physics*, 10(11):115021, 2008.

[122] G W Milton and S K Serkov. Neutral coated inclusions in conductivity and anti–plane elasticity. *Proceedings of the Royal Society of London A: Mathematical, Physical and Engineering Sciences*, 457(2012):1973–1997, 2001.

[123] G W Milton and J R Willis. On modifications of Newton's second law and linear continuum elastodynamics. *Proceedings of the Royal Society A: Mathematical, Physical and Engineering Science*, 463(2079):855–880, 2007.

[124] G S Mishuris, A B Movchan, and L I Slepyan. Localised knife waves in a structured interface. *Journal of the Mechanics and Physics of Solids*, 57(12):1958–1979, 2009.

[125] D Misseroni, D J Colquitt, A B Movchan, N V Movchan, and I S Jones. Cymatics for the cloaking of flexural vibrations in a structured plate. *Scientific Reports*, 6:23929, 2016.

[126] A B Movchan and S R L Guenneau. Split-ring resonators and localized modes. *Physical Review B*, 70:125116, 2004.

[127] A B Movchan and N V Movchan. *Mathematical Modelling of Solids with Nonregular Boundaries*. CRC Press, Boca Raton, FL, 1995.

[128] A B Movchan, N V Movchan, and C G Poulton. *Asymptotic Models of Fields in Dilute and Densely Packed Composites*. Imperial College Press, London, 2002.

[129] A B Movchan, N A Nicorovici, and R C McPhedran. Green's tensors and lattice sums for electrostatics and elastodynamics. *Proceedings of the Royal Society A: Mathematical, Physical and Engineering Sciences*, 453:643–662, 1997.

[130] A B Movchan, N A Nicorovici, and R C McPhedran. Green's tensors and lattice sums for electrostatics and elastodynamics. *Proceedings of the Royal Society of London A: Mathematical, Physical and Engineering Sciences*, 453(1958):643–662, 1997.

[131] A B Movchan and L I Slepyan. Band gap Green's functions and localized oscillations. *Proceedings of the Royal Society A: Mathematical, Physical and Engineering Sciences*, 463(2086):2709–2727, 2007.

[132] A B Movchan, V V Zalipaev, and N V Movchan. Photonic band gaps for fields in continuous and lattice structures. In *IUTAM Symposium on Analytical and Computational Fracture Mechanics of Non-Homogeneous Materials*, pages 437–446. Springer, 2002.

[133] A B Movchan, V V Zalipaev, and N V Movchan. Photonic band gaps for fields in continuous and lattice structures. In B L Karihaloo, editor, *IUTAM Symposium on Analytical and Computational Fracture Mechanics of Non-Homogeneous Materials: Proceedings of the IUTAM Symposium held in Cardiff, U.K., 18–22 June 2001*, pages 437–446. Springer Netherlands, Dordrecht, 2002.

[134] M J Musgrave. *Crystal Acoustics*, volume 184. Holden-Day, San Francisco, 1970.

[135] I Newton. *Philosophiæ Naturalis Principia Mathematica*. The Royal Society, London, 1687.

[136] N A Nicorovici, R C McPhedran, and G W Milton. Transport properties of a three-phase composite material: the square array of coated cylinders. *Proceedings of the Royal Society of London A: Mathematical, Physical and Engineering Sciences*, 442(1916):599–620, 1993.

[137] N A Nicorovici, R C McPhedran, and G W Milton. Optical and dielectric properties of partially resonant composites. *Physical Review B*, 49(12):8479, 1994.

[138] M J Nieves, A B Movchan, I S Jones, and G S Mishuris. Propagation of slepyan's crack in a non-uniform elastic lattice. *Journal of the Mechanics and Physics of Solids*, 61(6):1464–1488, 2013.

[139] E Nolde, R V Craster, and J Kaplunov. High frequency homogenization for structural mechanics. *Journal of the Mechanics and Physics of Solids*, 59(3):651–671, 2011.

[140] A N Norris. A theory of pulse propagation in anisotropic elastic solids. *Wave Motion*, 9(6):509–532, 1987.

[141] A N Norris. Acoustic cloaking theory. *Proceedings of the Royal Society A: Mathematical, Physical and Engineering Science*, 464(2097):2411–2434, 2008.

[142] A N Norris and W J Parnell. Hyperelastic cloaking theory: transformation elasticity with pre-stressed solids. *Proceedings of the Royal Society A: Mathematical, Physical and Engineering Science*, 468(2146):2881–2903, 2012.

[143] A N Norris and A L Shuvalov. Elastic cloaking theory. *Wave Motion*, 48(6):525–538, 2011.

[144] A N Norris and G R Wickham. Elastic waves in inhomogeneously oriented anisotropic materials. *Wave Motion*, 33(1):97–107, 2001.

[145] J F Nye. *Natural Focusing and Fine Structure of Light*. Institute of Physics Publishing, London, 1st edition, 1999.

[146] J A Ogilvy. Ultrasonic beam profiles and beam propagation in an austenitic weld using a theoretical ray tracing model. *Ultrasonics*, 24(6):337–347, 1986.

[147] F W J Olver. *Asymptotics and Special Functions*. Academic Press, New York, 1974.

[148] F W J Olver, D W Lozier, R F Boisvert, and C W Clark. *NIST Handbook of Mathematical Functions*. Cambridge University Press, Cambridge, 2010.

[149] J O'Neill, Ö Selsil, R C McPhedran, A B Movchan, and N V Movchan. Active cloaking of inclusions for flexural waves in thin elastic plates. *The Quarterly Journal of Mechanics and Applied Mathematics*, 68(3):263–288, 2015.

[150] J O'Neill, Ö Selsil, R C McPhedran, A B Movchan, N V Movchan, and C Henderson-Moggach. Active cloaking of resonant coated inclusions for waves in membranes and Kirchhoff plates. *The Quarterly Journal of Mechanics and Applied Mathematics*, 69(2):115–159, 2016.

[151] G Osharovich, M Ayzenberg-Stepanenko, and O Tsareva. Wave propagation in elastic lattices subjected to a local harmonic loading. II. Two-dimensional problems. *Continuum Mechanics and Thermodynamics*, 22(6-8):599–616, 2010.

[152] G G Osharovich and M V Ayzenberg-Stepanenko. Wave localization in stratified square-cell lattices: The antiplane problem. *Journal of Sound and Vibration*, 331(6):1378–1397, 2012.

[153] W M Ostachowicz and M Krawczuk. Analysis of the effect of cracks on the natural frequencies of a cantilever beam. *Journal of Sound and Vibration*, 150(2):191–201, 1991.

[154] G P Panasenko. *Multi-Scale Modelling for Structures and Composites*. Springer, The Netherlands, 2005.

[155] W J Parnell. Nonlinear pre-stress for cloaking from antiplane elastic waves. *Proceedings of the Royal Society of London A: Mathematical, Physical and Engineering Sciences*, 468(2138):563–580, 2012.

[156] W J Parnell, A N Norris, and T Shearer. Employing pre-stress to generate finite cloaks for antiplane elastic waves. *Applied Physics Letters*, 100(17):171907, 2012.

[157] W J Parnell and T Shearer. Antiplane elastic wave cloaking using metamaterials, homogenization and hyperelasticity. *Wave Motion*, 50(7):1140–1152, 2013.

[158] J B Pendry, D Schurig, and D R Smith. Controlling electromagnetic fields. *Science*, 312(5781):1780–1782, 2006.

[159] R Penrose. A generalized inverse for matrices. *Mathematical Proceedings of the Cambridge Philosophical Society*, 51(03):406–413, 1955.

[160] S B Platts and N V Movchan. Low frequency band gaps and localised modes for arrays of coated inclusions. In A B Movchan, editor, *IUTAM Symposium on Asymptotics, Singularities and Homogenisation in Problems of Mechanics*, pages 63–71, New York, 2003. Kluwer Academic Publishers.

[161] S B Platts and N V Movchan. Phononic band gap properties of doubly periodic arrays of coated inclusions. In D J Bergman and E Inan, editors, *Continuum Models and Discrete Systems*, pages 287–294. Springer, Dordrecht, 2004.

[162] S B Platts, N V Movchan, R C McPhedran, and A B Movchan. Two-dimensional phononic crystals and scattering of elastic waves by an array of voids. *Proceedings of the Royal Society of London A: Mathematical, Physical and Engineering Sciences*, 458:2327–2347, 2002.

[163] S B Platts, N V Movchan, R C McPhedran, and A B Movchan. Band gaps and elastic waves in disordered stacks: normal incidence. *Proceedings of the Royal Society of London A: Mathematical, Physical and Engineering Sciences*, 459:221–240, 2003.

[164] S B Platts, N V Movchan, R C McPhedran, and A B Movchan. Transmission and polarization of elastic waves in irregular structures. *Journal of Engineering Materials and Technology*, 125:2–6, 2003.

[165] A Poddubny, I Iorsh, P Belov, and Y Kivshar. Hyperbolic metamaterials. *Nature Photonics*, 7(12):948–957, 2013.

[166] C G Poulton, A B Movchan, R C McPhedran, N A Nicorovici, and Y A Antipov. Eigenvalue problems for doubly periodic elastic structures and phononic band gaps. *Proceedings of the Royal Society A: Mathematical, Physical and Engineering Sciences*, 456(2002):2543–2559, 2000.

[167] A P Prudnikov, Yu A Brychkov, and O I Marichev. *Integrals and Series*, volume 4. Gordon and Breach Science Publishers, Amsterdam, 1992.

[168] M Rahm, D Schurig, D A Roberts, S A Cummer, D R Smith, and J B Pendry. Design of electromagnetic cloaks and concentrators using form-invariant coordinate transformations of Maxwell's equations. *Photonics and Nanostructures-Fundamentals and Applications*, 6(1):87–95, 2008.

[169] Lord Rayleigh. On the influence of obstacles arranged in rectangular order upon the properties of a medium. *Philosophical Magazine*, 34:481–502, 1892.

[170] M Ruzzene, F Scarpa, and F Soranna. Wave beaming effects in two-dimensional cellular structures. *Smart Materials and Structures*, 12:363–372, 2003.

[171] M Saigo and H M Srivastava. The behavior of the zero-balanced hypergeometric series $_pF_{p-1}$ near the boundary of its convergence region. *Proceedings of the American Mathematical Society*, 110(1):pp. 71–76, 1990.

[172] E Sánchez-Palencia. *Non-homogeneous Media and Vibration theory*. Springer-Verlag, Berlin, 1980.

[173] D Schurig, J J Mock, B J Justice, S A Cummer, J B Pendry, A F Starr, and D R Smith. Metamaterial electromagnetic cloak at microwave frequencies. *Science*, 314(5801):977–980, 2006.

[174] D Schurig, J B Pendry, and D R Smith. Calculation of material properties and ray tracing in transformation media. *Optics Express*, 14(21):9794–9804, 2006.

[175] V M Shalaev. Transforming light. *Science*, 322(5900):384–386, 2008.

[176] R A Shelby, D R Smith, and S Schultz. Experimental verification of a negative index of refraction. *Science*, 292(5514):77–79, 2001.

[177] L I Slepyan. Dynamics of a crack in a lattice. *Soviet Physics–Doklady*, 26:538–540, 1981.

[178] L I Slepyan. *Models and Phenomena in Fracture Mechanics*. Springer-Verlag, Berlin, 2002.

[179] L I Slepyan and M V Ayzenverg-Stepanenko. Some surprising phenomena in weak-bond fracture of a triangular lattice. *Journal of The Mechanics and Physics of Solids*, 50(8):1591–1625, 2002.

[180] A Srivastava. Elastic metamaterials and dynamic homogenization: a review. *International Journal of Smart and Nano Materials*, 6(1):41–60, 2015.

[181] N Stenger, M Wilhelm, and M Wegener. Experiments on elastic cloaking in thin plates. *Physical Review Letters*, 108(1):14301, 2012.

[182] S P Timoshemko and J N Goodier. *Theory of Elasticity*. McGraw-Hill, New York, 1970.

[183] S P Timoshenko and S Woinowsky-Krieger. *Theory of Plates and Shells*, volume 2. McGraw-Hill, New York, 1959.

[184] B van der Pol and H Bremmer. *Operational Calculus Based on the Two-Sided Laplace Transform*. Cambridge University Press, London, 1950.

[185] J Wauer. Modelling and formulation of equations of motion for cracked rotating shafts. *International Journal of Solids and Structures*, 26(8):901–914, 1990.

[186] G R Wickham and I D Abrahams. The refraction of ultrasound in grainy and inhomogeneous fibre-reinforced materials. *Proceedings of the Institute of Acoustics*, 13:125–136, 1991.

[187] J P Wolfe. *Imaging Phonons*. Cambridge University Press, Cambridge, 1st edition, 1998.

[188] V V Zalipaev, A B Movchan, and I S Jones. Two-parameter asymptotic approximations in the analysis of a thin solid fixed on a small part of its boundary. *The Quarterly Journal of Mechanics and Applied Mathematics*, 60(4):457–471, 2007.

[189] V V Zalipaev, A B Movchan, and I S Jones. Waves in lattices with imperfect junctions and localized defect modes. *Proceedings of the Royal Society of London A: Mathematical, Physical and Engineering Sciences*, 464(2096):2037–2054, 2008.

[190] V V Zalipaev, A B Movchan, C G Poulton, and R C McPhedran. Elastic waves and homogenization in oblique periodic structures. *Proceedings of the Royal Society A: Mathematical, Physical and Engineering Science*, 458:1887–1912, 2002.

[191] I J Zucker. 70+ Years of the Watson Integrals. *Journal of Statistical Physics*, 145(3):591–612, 2011.

Index

Asymptotic
approximation, *see* Asymptotic Analysis
estimate, *see* Asymptotic Analysis
expansion, *see* Asymptotic Analysis
Asymptotic Analysis, 9, 11, 13, 14, 17, 20, 22, 37, 38, 41–44, 46, 47, 50, 53, 55, 59–62, 66, 68, 71–75, 82, 85, 87, 88, 93, 97–100, 115–117, 120, 123–130, 142, 166, 177, 183, 184, 186, 187, 191, 193, 196

Band Gaps, 2, 9, 14, 20–22, 28, 29, 31, 33–37, 43–47, 50–52, 55, 56, 59, 79–83, 90, 93, 95, 96, 104–107, 115, 117, 118, 121, 122, 124, 126, 206, 219, 225
acoustic, 5, 56, 57
phononic, 2, 6
photonic, 2
Betti Formula, 66, 70, 203
Bloch Parameter, *see* Bloch Vector
Bloch Vector, 19, 24, 29, 51, 53, 56, 78, 81, 95, 134, 137, 211, 214
Bloch–Floquet, 13, 90, 222
condition, 23, 24, 51, 53, 56, 78, 85, 211
waves, 2, 6, 8–11, 18, 19, 22, 23, 25, 28, 29, 42, 49, 50, 55, 72–74, 78, 83, 89, 93, 97, 170, 206, 211, 214, 219, 220
Boundary Layer, 9, 61, 62, 64, 65, 68–72

Brillouin Zone, 19–21, 42, 79, 81, 82, 88, 89, 103, 104, 107

Chirality, 7, 12, 199, 208, 209, 211, 212, 214, 219
Cloaking, 2, 5–8, 11, 12, 141–151, 154, 155, 157, 158, 160, 161, 164, 166–178, 180, 182–186, 188–191, 194–196
active cloaks, 142
Cracks, *see* Defects

Defects, 2, 5, 7–11, 13, 28, 29, 49, 60–62, 70–73, 83, 85, 86, 101, 115, 116, 118–121, 124–126, 129, 133–137, 139, 141, 182, 206
Diffraction, 2, 3, 101, 103, 104, 108, 154, 155
Dispersion, 1, 2, 6–10, 12–14, 17, 18, 23, 25, 28, 31, 72, 73, 83, 89, 93, 95, 97, 103, 108, 111, 205, 206, 208, 211, 212, 214, 217, 219, 220
diagram, 18, 19, 21, 22, 24, 25, 29, 34, 43, 44, 50–52, 55–57, 73, 74, 78–81, 85, 88–90, 93, 95, 103, 135, 139, 205, 206, 214, 225
equation, 14, 16, 17, 19, 21, 24, 30, 32, 34, 39, 41–44, 83, 97, 104, 106, 109, 110, 117, 135, 136, 205, 212, 217, 219, 221
saddle point, 81, 82, 101, 103, 104, 106–108, 111–114, 225, 227
surfaces, 2, 3, 8, 9, 57, 79,

81–83, 93, 101, 102, 106, 109, 111–113, 115, 212, 214, 217, 219, 222, 225

Dynamic Anisotropy, 2, 3, 5, 8, 9, 11, 12, 29, 83, 95–98, 101, 113, 115

Elementary Cell, 8, 20, 23, 24, 35–38, 49–51, 53, 56, 73, 74, 78, 79, 85, 87, 89, 93, 96, 97, 102, 103, 106, 115, 158, 206, 208, 211, 212, 220, 222

Energy, 9, 25, 27–29, 59, 90, 203, 206

Filtering, 1, 3, 6–8, 11, 20, 29, 49, 50, 73, 83, 89, 109, 125, 199, 220

Floquet Waves, *see* Bloch–Floquet Waves

Floquet–Bloch Waves, *see* Bloch–Floquet Waves

Focussing, 2, 3, 8, 11, 72, 73, 80, 83, 85, 90, 93, 109, 114, 199, 220, 225, 227

Fourier Parameter, *see* Bloch Vector

Fourier Transform, 32, 33, 35, 102
 discrete, 96, 102, 109, 116, 117, 134, 221

Green's Function, 11, 96, 150, 175, 177, 178, 193, 222
 lattice, 9, 29, 42, 44, 95, 96, 101, 102, 115–118, 120, 121, 123, 124, 138

Group Velocity, 20, 59, 79, 81–83, 93, 98, 104, 106, 113, 214, 225
 effective, 20
 negative, 90, 93, 225

Homogenisation, 3, 5, 8, 20, 37, 41, 47, 79, 136, 138–140, 157, 214
 dynamic, 3, 8, 13, 37, 38, 41, 42, 46, 47

Interfaces, 1–3, 5, 7–9, 11–13, 23, 25–29, 49, 61, 72, 73, 80, 83, 84, 89, 90, 93, 109, 142, 144, 145, 148–150, 158, 166, 170, 171, 189, 195, 196, 199–207, 220, 222, 225, 227

Linear Water Waves, 1, 5, 13, 14, 16–18, 206

Localisation, 1–3, 8, 10–12, 25, 29–35, 49, 72, 73, 89, 95, 96, 98, 100, 101, 136, 139, 199
 defect modes, 9, 12, 13, 28, 29, 34, 37, 72, 95, 96, 101, 115, 116, 118, 119, 125–127, 129, 134–136, 139, 140, 170, 206
 dynamic, 10, 34, 46, 95
 localised waveforms, 1–3, 5, 8, 9, 11, 29–31, 34, 36, 37, 95–98, 101, 113, 115, 203

Long-wave Limit, 3, 8, 20, 79, 95, 113, 115, 214

Low-frequency Limit, 14, 23, 24, 72, 79, 80, 155, 164, 214, 217, 225

Metamaterial, 6, 8, 155, 220
 hyperbolic, 8, 9
 parabolic, 8

Multipole, 50, 119, 139, 141, 166, 187, 188, 190, 193, 194, 197

Negative Refraction, 3, 5, 8, 11, 49, 72, 73, 80, 82, 83, 89, 90, 93, 148–150, 199, 220

Periodic
 arrays, *see* Periodicity
 media, *see* Periodicity
 structure, *see* Periodicity

Periodicity, 2, 5–7, 9, 13, 18, 20–25, 28, 29, 34, 47, 49–51, 53, 55, 56, 72–75, 78, 79, 81, 83, 85, 88–90, 93, 95, 102, 206, 208, 211
 periodic function, 19, 34, 106

Phase Velocity, 17, 147

Index

Resonance, 1, 2, 8, 9, 19, 20, 37–40, 42–44, 47, 49, 50, 53, 55–57, 72–75, 79–82, 85, 88, 104, 105, 107–109, 111–114, 135, 136, 141, 142, 199, 205, 206

Resonators, 1, 3, 11, 49, 50, 53, 55–57, 59, 60, 72–74, 76, 78–83, 85, 87–90, 93, 141, 208, 209

Slowness Contours, 3, 57, 82, 98, 103, 104, 106, 111–115, 146, 147, 214

Slowness Vector, *see* Slowness Contours

Spectral Problem, 37, 38, 51, 78, 90, 119, 222

Standing Wave, *see* Resonance

Stop Band, *see* Band Gaps

Transmission, 5–7, 13, 20, 25, 27–29, 59, 61, 73, 90, 166, 170, 189, 199, 201–204, 206, 207, 222, 225

Transmission Matrix, 13, 26, 27

Trapped Modes, *see* Localisation

Unit Cell, *see* Elementary Cell

Vortex, *see* Chirality

Waveguide, 9, 93, 116, 134, 138, 139

Weight Function, 61, 62, 65, 66, 69